Vesselii

From Illusions to Reality

Time, Spacetime and
the Nature of Reality

 MINKOWSKI
Institute Press

Vesselin Petkov
Institute for Foundational Studies Hermann Minkowski
Montreal, Quebec, Canada
http://minkowskiinstitute.org/
vpetkov@minkowskiinstitute.org

Cover: The Golden Hinde. Vancouver Island. Keith Freeman photo
August, 2006 (http://en.wikipedia.org/wiki/File:Ghinde1.jpg)

ISBN: 978-1-927763-00-1 (softcover)
ISBN: 978-1-927763-01-8 (ebook)

Minkowski Institute Press
Montreal, Quebec, Canada
http://minkowskiinstitute.org/mip/

For information on all Minkowski Institute Press publications visit our
website at http://minkowskiinstitute.org/mip/books/

Dedicated to the great thinkers who succeeded in freeing us from persistent illusions and to all who examine the issues raised in this book carefully.

Preface

This book is the first volume in the new *Understanding Reality Series* of the *Minkowski Institute Press*. The idea of this series is to acquaint the general public with the scientific view of the world and particularly with those advancements in the sciences which reveal new and often counter-intuitive features of the world. This series is one of the ways through which the *Minkowski Institute Press* provides continual science education for the general public in order that scientific culture become an inseparable element of the common culture of every individual.

As the greatest mystery in the world is its very existence, the first volume is devoted to the nature of reality – what is it that exists and how does it exist? The book explores what fundamental physics tells us about the physical world and how the scientific picture of what exists often differs disturbingly from the "common sense" view based on the way our senses reflect the world. Centuries-old illusions are identified by showing that they contradict experimentally-confirmed results of modern physics, which clears the way toward deeper understanding of reality. Special attention is devoted to the greatest illusion that the world exists only at the present moment of time. This illusion has been realized by many great thinkers, but so far the human race has been unable to free itself from it, prompting Einstein to write this: "the distinction between past, present and future is only a stubbornly persistent illusion."

Getting rid of such stubbornly persistent illusions by open-mindedly examining the implications of modern physics for the physical world can help us rise above the fog of everyday life and see Nature the way she herself is.

Montreal, 7 September 2013   *Vesselin Petkov*

Acknowledgments

Despite my initial intention to have a very long Acknowledgments section, I soon realized that it is absolutely impossible to thank individually all colleagues, students, and friends with whom I discussed the wide range of issues presented in this book.

I am grateful to the students who attended my courses in three departments of Concordia University (Montreal, Canada) for their stimulating participation in the class discussions (see the Appendix for more information).

I would like to thank the colleagues who attended the three *International Conferences on the Nature and Ontology of Spacetime* (2004, 2006, 2008) organized by the *International Society for the Advanced Study of Spacetime* and held in Montreal. We had unforgettable time together.

My special thanks also go to students and colleagues who attended and took part in the discussions of my three talks at the *Montreal Inter-University Seminar on the History and Philosophy of Science* (2002–2010), for which I was responsible.

I cannot find words to express my gratitude to my wife Svetla and to our son Vesselin (Jr) for so many things – ranging from the endless discussions of many of the subjects covered in the book to their unconditional support. Thank you for everything.

In the last several years Svetla and I had numerous and interesting discussions with our friend Linda Cochrane. We would like to thank her for her helpful suggestions.

CONTENTS

1 WHAT TO TRUST TO AVOID BELIEVING IN ILLUSIONS

It is darkest under the lantern according to an ancient proverb. Like its different formulations, this proverb also has different interpretations, one of which captures the essence of the way we form our views of the world. Over the centuries we have learned that Nature has given us her deepest features as apparently self-evident phenomena which explains why we understand the most familiar phenomena the least. For example, we all believe we know what motion is, and do not bother to think even a bit about what seems to be a trivial concept to the overwhelming majority of us. But as the Eleatics argued twenty-five centuries ago and as we will see in the next chapters this concept is anything but trivial. The situation is the same with other apparently self-evident phenomena of the physical world such as gravitation, inertia, mass, space, and particularly time. While discussing the nature of time sixteen centuries ago, Saint Augustine eloquently expressed our illusory understanding of the most profound phenomena of the world: "What, then, is time? I know well enough what it is, provided that nobody asks me; but if I am asked what it is and try to explain, I am baffled" [26, p. 118]

The most familiar and apparently unquestionably obvious feature of the world – its very existence – turns out to be the greatest mystery. However, fortunately, our nature is such that that seemingly unquestionable obviousness does not last throughout our lives. In our intellectual development, we all reach a turning point when we start asking the perennial existential questions: "What is the world?"; "What am I?"; "What is the meaning of the existence of the world and myself?". The German philosopher Schopenhauer expressed this transition to enlightenment perhaps in the best way: "The lower a man stands in intellectual respects the less of a riddle does existence

1

seem to him... but, the clearer his consciousness becomes the more the problem grasps him in its greatness" (quoted by William James in *The Problem of Being* [2]).

Plato's Allegory of the Cave

Once we reach the moment of intellectual awakening we start realizing how illusory perhaps most of our views of the world are, and try to identify and get rid of any illusions. However, the history of our civilization has shown that identifying and freeing ourselves from illusions about the world is not an easy process. Since ancient times thinkers have been suspecting that what our senses tell us about the world might not necessarily reflect the world the way it itself is. Their unanimous advice on how we can recognize and abandon illusions is to deepen constantly our knowledge about the world through education and self-education. Perhaps the most famous example is Plato's allegory of the cave in his book *The Republic* in which he describes the life of the uneducated man in order to "compare our nature in respect of education and its lack" [3]. Prisoners who spent all their lives in a cave where they were restricted to see only the shadows of artificial objects shaped as humans and animals on the wall of the cave believe that what they see are the real things [3]: "Then in every way such prisoners would deem reality to be nothing else than the shadows of the artificial objects." But when a prisoner is freed (i.e. educated) and allowed to look toward the light, he would see the actual objects and would realize "that what he had seen before was all a cheat and an illusion."

In the beginning of 21st century everyone seems to be sufficiently educated (or self-educated thanks to the enormous information that is now easily available) to form an adequate view of the world. However, as far as the physical world is concerned, it appears only a small fraction of our civilization holds a view that is firmly based on reliable knowledge provided by modern physics. The huge majority seems to hold views which contradict scientific facts about the physical world deduced from the existing experimental evidence. I think the main reason for this disturbing situation is the issue of what constitutes reliable knowledge (i.e., scientific facts) and even whether such knowledge is at all possible. Unfortunately, some scientists appear to hold an unproductive view on the nature of scientific theories according to which scientific theories are only *descriptions* of physical phenomena and for this reason we cannot deduce any reliable knowledge about the world from them since we can describe the same phenomena by other theories which present a different picture of the world.

However, most scientists hold a realistic view on the nature of scientific theories according to which *scientific theories, whose predictions have been confirmed by experiment, adequately reflect those features of the world represented by the theories.* Perhaps the best and most recent proof of that is the hunt for the Higgs boson – if physicists were unsure that their experimentally-confirmed theories adequately represented elements of reality (in this case the Higgs boson), they would not invest such huge effort and funds to test that theoretical prediction. If the Higgs boson were not discovered, then the theory of elementary particles which does not predict its existence would again make a definite claim about the world – that such a particle (with the specified properties) does not exist.

Although such examples convincingly demonstrate that accepted scientific theories do reveal the true nature of those parts and features of the world that are represented by the theories, there does not exist a unanimous consensus in exactly what sense and to what extent our theories of the physical world provide true knowledge about it. The main reason appears to be the question of whether an experimentally-confirmed theory will forever remain a correct theory about the elements of the world which it represents, and in this sense it can be regarded as the *final* theory of those elements (which means that the knowledge about those elements of the world, provided by the theory, will never be challenged and will therefore be regarded as true knowledge about that part of the world). This question is obviously of crucial importance since only such *final theories about the elements of the world they represent* can provide true knowledge about those elements – e.g., the existence of the Higgs boson is an example of a piece of true knowledge and the prediction of its *existence* constitutes an element of a theory which is final in a sense that any new theory of elementary particles will not make the Higgs boson more existent.

Sometimes it is tempting to think that if one day an experimentally-confirmed scientific theory is replaced by a new theory, the old theory might be wrong and therefore its claims about the world might be questioned and refuted. Such a temptation should instantly evaporate when an undeniable meta-theoretical fact is taken into account – that *experiments do not contradict one another*[1]. Indeed, since a theory is regarded as experimentally-confirmed when its predictions are confirmed by experiment, no new theory can disprove the first one *in the domain where its predictions were confirmed*; otherwise the experiments that confirmed the predictions of the new theory would contradict the experiments that confirmed the predictions of the first theory (e.g., the experiments at the Large Hadron Collider which confirmed the existence of the Higgs boson would be contradicted by other experiments which would confirm the prediction of a new theory, disproving the first one, that such a particle did not exist).

Because experiments do not contradict one another, a theory will never be proved wrong in the domain where its predictions have been experimentally confirmed. In this sense, such a theory is a final theory about that domain and therefore provides true knowledge about it. As I am quite aware of how philosophers and perhaps even some scientists might react to the two statements in the previous sentence (on a final theory and true knowledge), let me give two examples to clarify the meaning of these statements.

- In addition to the example of the existence of the Higgs boson, the existence of the other elementary particles (electrons. protons, etc.) also constitutes pieces of true knowledge about

---

[1]A skeptic may, as always, object that *so far* we have not observed such contradictions, but that might happen in the future. By the same "logic," a skeptic may also say "So far men have not given birth, but that might happen in the future." I think the philosophical doctrine of skepticism is still around only due to its continued total ignorance of how science actually works. If skeptics try to understand that scientific statements are not based on *inductive inferences*, but reflect proper understanding of the mechanism of physical phenomena revealed by experimentally tested theories, they would realize why the *universality* of scientific truths (confirmed by experiment) cannot be questioned.

the world and in this sense the Standard Model is a final theory about the domain where its predictions of the *existence* of *those* elementary particles were experimentally confirmed. For this reason, we do possess true knowledge about the existence of the known elementary particles and, due to its being *true*, that knowledge is *final* – no future theory will disprove the Standard Model in the domain where its predictions of the known elementary particles were experimentally confirmed (since no future experiments will contradict the experiments which confirmed the *existence* of those particles).

- The predictions of classical mechanics and particularly Newton's three laws have been repeatedly confirmed by experiment and for this reason classical mechanics will never be proved wrong in its domain of applicability (where its predictions have been experimentally confirmed at velocities much smaller than the velocity of light and in the case of macroscopic bodies). The situation with Newton's gravitational theory is more complicated and the analysis of the experimental verification of its predictions requires special care in order to determine the theory's proper domain of applicability. For example, Newton's theory correctly predicts how bodies fall toward the Earth's surface, but interprets that fall as being caused by a gravitational force. However, according to the modern theory of gravitation – general relativity – falling bodies move by inertia since their fall is not caused by a gravitational force, but is a manifestation of the curvature of spacetime in the Earth's vicinity (induced by the Earth's mass). Newton's theory also correctly predicts that a body at rest on the Earth's surface is subject to a gravitational force (the body's weight), and general relativity did not question the existence of that force, but clarified its nature by showing that it is inertial, not gravitational. So general relativity did not (and could not) refute the predictions of Newton's gravitational theory which were confirmed by experiment – that (and how) (i) a body falls toward the Earth (and planets orbit the Sun while falling toward it), and (ii) a body prevented from falling (e.g. while being at rest on the Earth's surface) is subject to a force. General relativity provided deeper understanding of what causes those phenomena – the curvature of spacetime – as we will see in Chapter 6.

I think the following examples most convincingly demonstrate to people who do not have a professional science background that an experimentally-confirmed theory will never be proved wrong in its domain of applicability (that domain of the world where the theory's predictions have been repeatedly confirmed by experiment), and therefore provides true knowledge about that part of the world. A thousand years from now, it will still be classical mechanics – Newton's three laws and Newton's gravitational theory (for determining the weight of objects) – that will be used in the calculations when people build buildings and bridges, for example. A thousand years from now, it will still be classical (Maxwell's) electrodynamics that will be used in the calculations of electrical motors, the electrical wiring of buildings, ordinary electrical and microwave ovens (if such things will be used then), etc.

Scientists, particularly physicists, know that the very way science works is the best proof of the "eternity" of the experimentally-confirmed theories (as physics is the strongest example in this respect, when I talk about science in this book I mean physics). Indeed, the reliable scientific knowledge provided by such theories form the foundation on which new theories are built. Such foundational knowledge has no expiration date. In 1909 Max Planck explicitly stressed that foundational knowledge (whose elements he properly called invariants) plays a central role in the advancement of physics [5]:

> The principle of relativity holds, not only for processes in physics, but also for the physicist himself, in that a fixed system of physics exists in reality only for a given physicist and for a given time. But, as in the theory of relativity, there exist invariants in the system of physics: ideas and laws which retain their meaning for all investigators and for all times, and to discover these invariants is always the real endeavor of physical research. We shall work further in this direction in order to leave behind for our successors where possible – lasting results. For if, while engaged in body and mind in patient and often modest individual endeavor, one thought strengthens and supports us, it is this, that we in physics work, not for the day only and for immediate results, but, so to speak, for eternity.

Since the dawn of science in the 17th century (mostly due to the work of Galileo and Newton) until the two major revolutions in physics in the beginning of the 20th century (relativity and quantum mechanics), foundational knowledge (i.e. reliable scientific knowledge) about the world consisted mostly of assertions about the *existence* of physical phenomena such as:

- Inertia with its two aspects – (i) a free body moves by inertia, i.e. with constant velocity (constant speed and constant direction), and (ii) a body which is subject to a force (that prevents the body from moving by inertia) resists the change in its inertial motion with constant velocity (i.e. resists its acceleration). The second aspect of the phenomenon of inertia – resistance to acceleration – is captured in Newton's second law ($\mathbf{F} = m\mathbf{a}$) whose profound meaning is that in order to prevent a body from moving by inertia (i.e. from moving with constant velocity) a force must be applied to *overcome* the resistance which the body offers to its acceleration (i.e. to the change in its velocity).

- The equivalence of action and reaction in mechanical interactions as reflected in Newton's third law – if a body exerts a force on another body, the other body instantly reacts by exerting a force equal in magnitude and opposite in direction to the force exerted by the first body.

- All bodies fall toward the Earth with the same acceleration first realized and proved by Galileo; later Newton *interpreted* that experimental fact in terms of his second law of motion – as the falling bodies accelerate they should be subject to a force (the force of gravity) which accelerates them.

- A body at rest on the Earth's surface is subject to a force (the body's weight) regarded by Newton as gravitational.

These examples of pieces of foundational knowledge, provided by Newton's mechanics, effectively asserts only the existence of the phenomena reflected in those pieces of knowledge. No real attempt has been made to *explain* those phenomena – e.g., the resistance a body offers to its acceleration was *called* (not explained) 'the body's inertia'; the cause of a body's fall and its weight were merely *labeled* 'gravitational force' with no attempt to explain the nature of that 'force' (Newton himself explicitly stated that he only *described* gravitational phenomena and made no hypothesis on their nature; even he did not believe that the 'gravitational force' he introduced to describe the gravitational phenomena could be transmitted through the empty space separating celestial bodies).

What constitutes foundational knowledge dramatically changed after the advent of special and general relativity. Not only were new phenomena added to the existing foundational knowledge about the physical world (e.g. the equivalence of mass and energy, the relativistic mass increase, and the general relativistic frame dragging effect according to which a rotating body twists the surrounding spacetime), but also *explanations* of existing and relativistic phenomena became part of foundational knowledge. In Chapter 5 we will see why explanations, provided by the theory of relativity, do constitute foundational knowledge – *if those explanations were wrong, the experiments that confirmed the explained phenomena would be impossible.* We will have the most general proof of this strong statement shortly (in this chapter) and in detail in Chapter 5 – both the theory of relativity and the experiments which confirmed its kinematical predictions would be impossible if the world were *not* four-dimensional (with time as the fourth dimension as revealed by the theory of relativity), but three-dimensional (as our senses seem to imply). This proof is one of the best demonstrations that our theories (in this case the theory of relativity) provide true knowledge about the world.

As explanations of physical phenomena are an integral part of foundational knowledge, it is evident that only one theory of given physical phenomena provides their true explanation, which, in turn, demonstrates that scientific theories are something more than mere descriptions. Despite this, occasionally one can hear or read that different theories which describe the same physical phenomena are equivalent since they are just different descriptions of the phenomena. In some cases such a position is fine, but in other cases it is plain wrong which is best seen by the very fact of how physics works – that experiments are always performed to *choose* one of the *competing* theories describing the *same* physical phenomena.

Part of the art of doing physics is to determine whether different theories are indeed simply different descriptions of the same physical phenomena (as is the case with the three representations of classical mechanics – Newtonian, Lagrangian, and Hamiltonian), or *only one* of the theories competing to describe and explain given physical phenomena is the correct one (as is the case with general relativity, which identifies gravity with the non-Euclidean geometry of spacetime, and other alternative theories, which regard gravity as a force). The difference between these two cases can be illustrated by an example from everyday life. The first case – different theories are just different descriptions of a given phenomenon – is like the description of an event in different languages; every language correctly describes the event and therefore the different languages' accounts are equivalent. The second case – only one of the theories describing a given phenomenon is correct – is like different accounts of the same event (in any language) and, obviously, only one is correct.

The theory of relativity enriched the foundational knowledge of the physical world with explanations of physical phenomena. But, unfortunately, the other revolutionary theory of the 20th century – quantum mechanics – did not provide us with explanations of quantum phenomena that can be added to the foundational physical knowledge. Like in the case of the Newtonian mechanics, the foundational knowledge provided by quantum mechanics contained assertions only about the *existence* of new quantum phenomena. This is not surprising when it is taken into account that, like the Newtonian mechanics, which is the first experimentally-confirmed theory of the macroscopic world, quantum mechanics is the first experimentally-confirmed theory of the microscopic world (the quantum scale of the world). By contrast, the theory of relativity is the second theory of the macro scale of the world and as such is a better representation of the macro world, which enables it to reveal the true explanations of the represented phenomena.

In addition to augmenting the foundational knowledge of the physical world, the theory of relativity and quantum mechanics provided an important piece of reliable knowledge about how knowledge itself grows – new theories are more accurate representations of the world than the previous experimentally-confirmed theories and do not disprove them but incorporate them as limiting cases. After the advent of the theory of relativity and quantum mechanics this fact has been adopted by physicists as one of the necessary conditions which any new theory should meet – the predictions of any new theory should coincide with the predictions of the previous theory in the domain where those predictions have been tested by experiment.

How scientific knowledge grows by preserving the already accumulated foundational knowledge is nicely illustrated by an everyday example [4]:

> Quantum mechanics ... doesn't displace Newtonian mechanics, but incorporates it as a limit. Scientific theories grow by incorporating what is already known and adding to it, just as a tree adds layers on the outside while preserving its heartwood.

That accepted theories cannot be proved wrong in the domain where their predictions have been confirmed by experiment cannot be seriously questioned. But that some predictions of accepted theories contradict experimental results cannot be questioned either. It is precisely such contradictions that give rise to the occasional temptation to declare that if a prediction of a theory is contradicted by experiment, such a theory is wrong. Fortunately, science (physics in particular) does not work in that way at all. For example, the equations of motion of both special and general relativity spectacularly fail to describe the behaviour of quantum objects, but no one declares these theories wrong. Simply, in that case they have been employed outside of their domain of applicability where their predictions have been tested by experiment. For exactly the same reason some predictions of Newtonian mechanics contradict experimental results. Now we know that such contradictions do not prove that a theory, whose predictions have been experimentally confirmed, is wrong; they are an indication that the theory has been employed beyond its domain of applicability.

An example illustrates even better than the previous example why the already established foundational knowledge cannot be wrong. The top image on the right is a satellite picture of Montreal with low resolution. This picture can be thought of as an analog of Newtonian mechanics. There are a number of features of Montreal which are clearly seen on the satellite picture – the existence of recognizable streets and blocks of buildings especially in the downtown area. These features are the analog of foundational knowledge provided by Newtonian mechanics. No satellite pictures of increasing resolution will disprove what is clearly seen on the first picture; in exactly the same way, no theory will disprove Newtonian mechanics in the domain where its pre-

Satellite picture of Montreal with low resolution.

A higher resolution satellite picture of the region of Montreal shown in the bottom-left corner of the above picture.

dictions have been experimentally confirmed. However, on the low-resolution image there are areas whose features are not clearly distinguished – for example, the bottom-left corner of the picture does not provide unambiguous knowledge about the streets and blocks of buildings there. If we try to employ the knowledge of the downtown part of Montreal to that area, practically certainly we will fail as a new satellite picture with higher resolution (the bottom image) will show (or if we simply go to the area on the image). In the same way, theories employed outside of their domain of applicability fail. Evidently, the satellite picture with higher resolution can be thought of as an analog of a new theory (e.g., special relativity) describing the same 'part' of the world (the macroscopic scale), which is described by Newtonian mechanics.

In order that one forms an adequate view of the physical world based on scientific results, the issue of why a physical theory will never be proved wrong in its domain of applicability (where its predictions have been experimentally confirmed) should be thoroughly understood and not brushed aside as purely academic. Only then one can understand why scientific theories correctly represent those features of the world, which they describe. That is why, it should be stressed that the stakes in the question of whether or not scientific theories provide true knowledge of the world are at the highest possible level. The reason is that *an affirmative answer to this question will make us trust in even counter-intuitive features of the world revealed by scientific theories, whereas a negative answer would unavoidably imply that science tells us nothing about the world and the only information about it comes from our senses.*

To see even better why the stakes are really at the highest level as far as the question of the nature of scientific theories is concerned, let us face the fact that for centuries our senses (e.g., sight and touch) have been a continued source of illusions since they do not provide sufficient and unambiguous information that would allow us to form an adequate view of the physical world. An example dealing with the central question in the book – what is the nature of reality – is *whether the information coming from our senses enables us to determine how the world exists in time.* Through our senses we are aware *only at the present moment* of our own continued existence and the continued existence of the world and uncritically interpret these sense data to mean that we (as physical bodies) and the world itself exist only at the moment 'now', which constantly changes. In other words, for centuries we have been assuming without examination that

- the fact that we realize ourselves and the world only at the present moment implies that only this moment – the moment 'now' – exists,

- the fact that we realize the existence of the world only at the moment 'now' implies that the world *itself* exists solely at that moment, and

- the continued existence (endurance) of the world through time implies that the only existent present moment constantly changes, i.e. that time flows.

For centuries these three assumptions have been regarded as a self-evident foundation of a world view, now called *presentism*, according to which only the constantly changing moment 'now' exists and for this reason everything that exists, exists solely at this moment. Stated another way, on the presentist view, only the present, which since Aristotle (as shown in the next chapter) has been regarded as a three-dimensional world, exists, whereas the past does not exist any more and the future does not exist yet. The assumption that time flows embodies the very essence of the presentist view – the existent present (the three-dimensional world at the only existent present moment) constantly stops existing by becoming past and the non-existing future constantly comes into existence by becoming present.

What prompted the writing of this book was the fact that despite the overwhelming evidence against presentism, this view continues to be held by the huge majority of our civilization even in the 21st century. In the next chapter we will see that since ancient times there have existed logical arguments against the sole existence of the present moment. In Chapter 4 and especially in Chapter 5 we will discuss in detail how the theory of relativity decisively disproved the presentist view of the world. Now I will only summarize some of the facts which (i) demonstrate the inadequacy of presentism due to its being based on information coming from our senses, which has never been properly examined by those sharing this view, and (ii) demonstrate how presentism contradicts both the theory of relativity and the experiments which confirmed its predictions.

Let us start by asking what the evidence for the presentist view is. Throughout the centuries, up to this moment, the only "evidence" has been the apparently self-evident interpretation of our sense data captured in the above three implicit assumptions behind presentism. Nothing else, no experimental evidence coming either from classical or modern science. Although it is clear, but since this is about our basic view of the world, let me specifically ask you not to believe the statement that there is no experimental evidence for presentism without reading opposite views. I think the best and fair approach is to make your own decision only after having read, in addition to this book, the most recent defence of the key feature of presentism (the reality of time flow) by several well-known physicists [6]-[9]. Here I will only comment on what George Ellis wrote in December 2008 when we discussed the reality of the flow of time while participating in the 2008 Essay Contest *The Nature of Time* organized by the Foundational Questions Institute (FQXi) [11]: "the real-world evidence that time does indeed flow is overwhelming (example: this posting was not posted till I posted it at a particular proper time along my world line)", and his response to my insistence that our everyday experience by itself (without being critically examined) does not constitute an argument for the reality of the flow of time: "I totally disagree. Our daily life experience is abundant evidence about the nature of reality. If physicists chose to ignore all that evidence, then their theories are not adequately related to the real world. They are trapped in the 'isolated laboratory' view of physics, a convenient fiction (no truly isolated laboratories exist either in space or time), instead of believing that physics should be able to describe the real world, as I do."

I think Ellis' comments eloquently show what a stubbornly persistent illusion the flow of time (which is behind the presentist view) indeed is. The apparently self-evident interpretation that our daily life experience (reflected in our sense data) proves the reality of time flow could lure even renowned scientists, like Ellis, who believe that physics does describe the real world and, more inexplicably, who are not afraid to correct their views if confronted with convincing arguments. Let us examine Ellis' comments to see that they contain no evidence whatsoever for the key element of the presentist view – that time really flows (Ellis believes that the time flow is real and holds a view according to which both the past and the present are real, but the future is not).

For Ellis "Our daily life experience is abundant evidence about the nature of reality" and is therefore evidence about the reality of time flow. Undoubtedly, he means *experimental* evidence since everything we observe in our daily life is indeed an enormous set of natural experimental facts – for example, bodies staying at rest, moving and colliding. Ellis certainly regards as such natural experimental evidence also the facts that we are aware of the existence of the world (and of our own existence) only at the present moment and for this reason we *do not observe* the future, and asserts that all these facts observed in our everyday life prove the reality of time flow and particularly his view that only the future does not exist but part of it constantly comes into existence while becoming present (this transition of the non-existing future events into the existing present events is precisely what is regarded as a real flow of time).

Ellis' comments do not provide any evidence for the reality of time flow since they contain the same implicit assumptions behind the belief that time flow is real (listed above), all of which turn out to have no experimental support. He almost explicitly takes for granted what must be proved – that the future does not exist (which could be a proof for the reality of time flow). This is seen in his own example above – "example: this posting was not posted till I posted it at a particular proper time along my world line". That is, he merely states (instead of proving) that the event of posting his comments did not exist when it was in the future and came into being when it became present. Like all who believe that this is true, i.e. that time really flows, Ellis *just asserts* that "the real-world evidence that time does indeed flow is overwhelming".

Before showing in more detail why what Ellis calls "the real-world evidence" does not constitute experimental evidence in the proper scientific sense, let me present a more general argument which, taken even alone, is sufficient to prove that. As we will see shortly and in great detail in Chapter 5, the *experiments* which confirmed the kinematical relativistic effects proved that past, present and future exist equally. As experiments do not contradict one another "the real-world evidence" cannot be regarded as experimental evidence for the non-existence of the future; otherwise "the real-world evidence" would contradict the relativistic experiments. In this situation, it is evident that the scientific approach available to all who continue to insist that the flow of time is real is to refute the assertion that the relativistic experiments proved the equal existence of all moments of time.

Let us now see why the apparently self-evident interpretation of our sense data coming from our daily life experience (which gives rise to the illusion that time flows and to the presentist view) turns out to be a misinterpretation. This becomes immediately evident when we start asking questions to determine whether the information coming from our senses is unambiguous.

First, the fact that we realize ourselves at one single moment of time is not a proof at all that only that moment exists. To understand better the huge logical jump we make, without realizing it, when we take it as self-evident that only 'now' exists since we are aware only of this moment, consider the analogous situation with space. We are aware only of a relatively small spatial region around us, but we do not claim that only that region of space exists. It is clear why we do this – we have different kinds of indications of the existence of the rest of the space and despite that we are not directly aware of the whole of space we are certain that it exists. But we have similar indications about time – we are aware that the past moments of time existed (like distant regions of space which we visited in the past and saw that they existed), and we are certain that there will be future like there was past. Now, some may be tempted to argue that we are aware of whole regions of space *at once*, but we are always aware of a *single* moment of time – the moment 'now'. I think it will be a good idea to suppress such a temptation because it is another illusion that we are aware of whole regions of space *at once* as we will see below. Even if we forget the analogy with space, there exist three arguments against the sole existence of the present moment (and the third is decisive) – (i) the assumption that only 'now' exists leads to logical contradictions realized even by some ancient thinkers as we will see in the next chapter, (ii) there does not exist anything that even resembles a piece of physical experimental evidence for the sole reality of the present moment; if only 'now' existed, physics would have proven it by now, and (iii) we will see in Chapter 5 that the experiments which confirmed the theory of relativity provided the most convincing proof for the equal existence of all moments of time (because the theory of relativity would be impossible if only the present moment existed).

Second, the fact that we realize the world at one single moment of time is not a proof at all that the world itself exists only at that moment. We do *see* the world around us only at the present moment, but what we see tells us nothing definite about the world. When we believe we see the world at the moment 'now', we are actually aware at this moment of mental images of different objects, which contain information coming from our senses. Even a quick analysis reveals that we need extra information in order to understand what those images represent. For example, we believe that the mental images in our mind represent three-dimensional objects, but these same images could also represent three-dimensional reflections from extra-dimensional objects. Sitting in our armchairs and contemplating about the two options will never allow us to determine which of them is the true one. Before 1908, when Einstein's mathematics professor Hermann Minkowski showed that Einstein's theory of special relativity is in fact a theory of a four-dimensional world (with all moments of time forming the fourth dimension), we had been uncritically interpreting the mental images to mean that they represent three-dimensional objects and a three-dimensional world. After 1908, the relativistic experimental evidence has been gradually convincing physicists and philosophers that the counter-intuitive interpretation of the mental images in our mind (that those images represent three-dimensional reflections from a four-dimensional world) is the correct one; indeed, as we will see shortly (and in greater detail in Chapter 5) the experiments which confirmed the kinematical predictions of special relativity proved the reality of Minkowski's four-dimensional world, which we now call spacetime (or sometimes block universe since spacetime is actually the whole history of the Universe given *en bloc*).

Rømer's discovery in 1676 that the speed of light is finite provided another example that additional information is needed to determine how to interpret the mental images through which we are aware of the external world. Before that discovery people believed that what they see at the present moment exists at that moment, but it turned out that that first naive form of presentism was an illusion – the world we see at the moment 'now' does not exist at the moment we perceive it because due to the finite speed of light we see only *past* events. When we look at a cloudless night sky we see the Moon and a myriad of stars and tend to believe that all of them and the whole world exist at once right now – at the moment 'now' when we perceive them. However, this is an obvious illusion – the mental image of the Moon of which we are aware at the present moment represents the image of the Moon which is about 1.3 seconds old (as the average distance between the Moon and the Earth is about 400,000 km and the speed of light is 300,000 km/s, it takes about 1.3 seconds for the sunlight reflected from the Moon's surface to reach our eyes), whereas the mental images of some stars may represent images of those stars that are millions of years in the past. Therefore, the mental images of which we are aware *at once at the present moment* represent images of objects that *existed at different moments in the past*. In other words, our feeling that we are aware at the moment 'now' of whole regions of space *at once* also turns out to be an illusion – space is defined in terms of *simultaneity* (as all space points corresponding to the *same* moment of time), whereas the space region we 'see' (through the distances between the objects in it) *at once* at the moment 'now' does not constitute a space region at all since it is a mosaic of small space fractions corresponding to *different* past moments of time (the greater the distance of a space fraction from us is, the more in the past it is since it takes more time for light reflected from the objects in that space fraction to reach us).

So, rigorously speaking, our senses do not tell us anything definite about what the world itself is. Only science can provide additional information which can enable us to understand what the sense data represent. That is why, in order that we hold an adequate view of the world (especially if it turns out to be frighteningly counter-intuitive), it is of paramount importance to have proof that scientific theories provide true knowledge of the world. Fortunately, as we saw above fundamental physical theories give the best proof for that. Perhaps the most spectacular proof of true knowledge of the world comes from the theory of relativity and particularly from the experiments which confirmed its kinematical predictions. That proof is spectacular on two counts. First, a specific scientific theory radically affects our view of the world by disproving the presentist view and showing that the world is four-dimensional with time as the fourth dimension. Second, how powerful this proof of true knowledge of the world is can be demonstrated by assuming for a moment that the knowledge deduced from the theory of relativity and the experiments which confirmed it is not true knowledge, i.e. that the world is three-dimensional and evolves as time really flows. Then neither the theory of relativity nor the relativistic experiments would be possible, if the world were three-dimensional. Let us sketch that proof now and we will return to it in Chapter 5.

After Rømer's discovery of the finite speed of light showed that the naive version of presentism (what we *see simultaneously* 'now' exists at this moment) was an illusion, the presentist view had been silently adjusted to accommodate that scientific fact. Its modified version is still the world view shared by the huge majority of our civilization in the 21st century which is truly inexplicable given the fact that presentism openly contradicts the relativistic experimental evidence. What makes presentism easily tested experimentally is that it is defined in terms of *simultaneity* – the present is the three-dimensional world (the space and all objects in it) which exists solely at the present moment, i.e. the present is everything that exists *simultaneously* at the moment 'now' (we saw above that space is defined in terms of simultaneity – as all space points corresponding to the *same* moment of time). As seen in the figure, the present is a class of simultaneous events (i.e., the class of all objects and space points which exist simultaneously at the present moment, since an *'event'* in the theory of relativity means 'an object or a space point at a given moment of time').

In 1905 Einstein formulated his special theory of relativity whose major result was that *observers in relative uniform motion have different times*. That was the end of a very stubbornly persistent illusion – that there existed one absolute time and the whole world was subject to its flow. The very essence of the presentist view is entirely based on the idea of absolute time – there is one absolute 'now' for the whole world and that is why, on the presentist view, everything that exists is

18

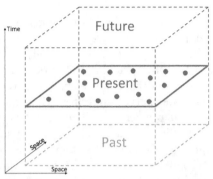

On the presentist view only the present – the three-dimensional world at the moment 'now' – exists. The past does not exist any more and the future does not exist yet.

regarded as existing simultaneously at the present moment. It seems logically unavoidable that getting rid of the illusion of absolute time should have immediately led to getting rid of the presentist illusion since according to special relativity observers in relative motion have different times, different nows, and different classes of simultaneous events, which means they have *different presents* (i.e., different three-dimensional worlds). What looked logically unavoidable, however, has not happened for over a century despite that in 1908 Minkowski announced a stunning new world view rigorously following from the fact that the many times of many observers in relative motion (introduced by Einstein) imply many spaces as well, which is *not possible in a three-dimensional world* and therefore unavoidably leads to the conclusion that what exists is an absolute (the same for all observers) four-dimensional world [12, p. 114]:

> Hereafter we would then have in the world no more *the* space, but an infinite number of spaces analogously as there is an infinite number of planes in three-dimensional space. Three-dimensional geometry becomes a chapter in four-dimensional physics.

Minkowski's argument that many spaces imply a four-dimensional world in which all moments of time form the fourth dimension is shown in the figure where the cube represents such a four-dimensional world (spacetime). Two observers in relative motion have different spaces which are represented in the figure by the horizontal and the inclined surfaces. The spacetime diagram makes Minkowski's argument completely obvious – the consequence of special relativity that two observers in relative motion have different spaces (first realized by Minkowski) implies that the world is four-dimensional because only then the two observers can have different spaces which are simply two three-dimensional cross sections of spacetime. Minkowski's explanation that observers in relative motion can have different spaces and times only in (at least) a four-dimensional world removed the mystery around Einstein's initial formulation of special relativity where Einstein merely postulated that time and simultaneity were relative. The spacetime diagram naturally explains the deep physical meaning of Einstein's relativity of simultaneity – the two surfaces (horizontal and inclined), which represent the spaces of the observers, also represent the observers' different classes of simultaneous events (since a space is a class of simultaneous events); therefore relativity of simultaneity clearly implies that the world is four-dimensional.

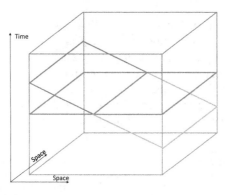

Two observers in relative motion can have their own relative spaces only if the two spaces are cross sections of an absolute underlying reality – spacetime.

There is some irony in the discovery of special relativity. As we will see in Chapter 4, Einstein wanted to get rid of all absolutes in physics – absolute motion, absolute space, absolute time, and absolute simultaneity, and postulated that they are all relative. This seems to have been so important to him that he named his theory published in 1905 'the theory of relativity'. Einstein's theory successfully explained the existing experimental evidence and made new radical predictions which were confirmed by experiment. Despite this success, the physical meaning of the *relativity* of physical quantities remained a complete mystery until 1908 when Minkowski showed that he could still have an important lesson to teach his already famous student. Minkowski arrived at the revolutionary view that space and time form an inseparable *absolute* four-dimensional world by realizing that relative space, relative time, and relative simultaneity can exist only in such an absolute world (as we saw in the figure above). The essence of Minkowski's lesson is that *relative quantities are manifestation of an underlying ab-*

solute reality.

Minkowski convincingly showed that merely postulating that motion, space, time, and simultaneity are relative, in order to explain the experimental evidence, is only the first step toward the true explanation (which should involve some absolute reality) of that experimental evidence since *no relative quantities would be possible if no underlying absolute reality existed.* It is this fact which made perhaps the greatest and most spectacular proof in the history of science possible – the theory of relativity and the relativistic experimental evidence *proved* that the world is four-dimensional; otherwise, if the world were three-dimensional, the theory of relativity and most importantly the experiments which confirmed its kinematical predictions would be impossible.[2]

This is clearly seen in the spacetime diagram above. If the world were three-dimensional, the cube in the diagram (representing spacetime) would be reduced to the horizontal plane (which represents a three-dimensional world, i.e. the present) and all observers in relative motion would share the *same* three-dimensional world, i.e. the *same* space and the *same* class of simultaneous events; therefore, space, time, and simultaneity would be *absolute* in contradiction with the theory of relativity.

---

[2]I am aware only of a single criticism of the arguments that the relativistic kinematical effects are impossible in a three-dimensional world, which are mere expansion of Minkowski's own arguments that the theory of relativity is a theory of a four-dimensional world. In a book that appeared this year (2013) F. Weinert took the *conclusion* of rigorous analyses of the relativistic effects (including an analysis of Minkowski's explanation of length contraction) to be the *argument* and tried to construct a logical fallacy. He claimed that the "argument" (if the world is four-dimensional, then we observe relativistic effects) affirms the consequent (the world is a four-dimensional block universe). What Weinert regards as the argument (or rather the premise!) is not the argument; it is the conclusion deduced from analyses of the relativistic effects and the experiments that confirmed them (as outlined here and discussed in Chapter 5). Therefore, the statement "the world should be four-dimensional in order that we observe relativistic effects" is supported by experiment (the essence of my argument, not touched by Weinert) and for this reason what Weinert constructed is not a logical fallacy, but a perfect argument. I doubt that such kind of criticism, without even touching the real arguments and analyses, can foster cooperation between scientists and philosophers. I will mention only two other wrong claims in his book – (i) that "Minkowski himself did not necessarily accept the notion of the block universe" (p. 138), and (ii) that "in his later years Einstein wavered in his support for the Parmenidean view" ("the Parmenidean block universe") (p. 141). In both cases, the reasons given by Weinert are perfectly explainable in terms of spacetime and both Minkowski and Einstein knew that. The wrong claim regarding Einstein is particularly inexplicable because Weinert undoubtedly knows well that Einstein expressed his strongest support for the block universe a month before he left this world: "the distinction between past, present and future is only a stubbornly persistent illusion" [15].

If some suspect that the contradiction is implied by the space-time diagram itself, then simply forget the diagram. If reality were a three-dimensional world, simultaneity would be absolute in contradiction with relativity since *a three-dimensional world constitutes a single class of simultaneous events* – everything that exists *simultaneously* at the present moment (if now some suspect that the three-dimensional world is defined in such a way which leads to the contradiction with relativity, try to define it differently).

I hope you now see why I wrote that the theory of relativity's proof of the reality of Minkowski's absolute four-dimensional world is the greatest and most spectacular proof in the history of science. Never before has a scientific theory so decisively and unambiguously disproved a world view and proved another as the theory of relativity did.

The presented here sketch of the proof of the reality of spacetime (i.e. that the world is four-dimensional) involves only the major result of the theory of relativity – relativity of simultaneity (i.e., the fact that observers in relative motion have different times and different spaces). To make the proof even more convincing two remarks should be made.

First, relativity of simultaneity has never been directly tested experimentally. However, two relativistic effects – length contraction and time dilation – are specific manifestations of relativity of simultaneity and these effects have been confirmed by experiment as we will see in Chapter 5. Therefore, not only would relativity of simultaneity be impossible if the world were three-dimensional, but, what is crucial, *the experiments*, which confirmed length contraction and time dilation, would be impossible in a three-dimensional world.

Second, rigorously speaking, Minkowski's proof that the world is four-dimensional is valid only if the *existence* of physical objects and the world itself is *absolute*; only then it is possible for two observers in relative motion to have different spaces – the *relative* spaces of the two observers, represented by the horizontal and the inclined surfaces in the spacetime diagram above, are manifestations of an *absolute* four-dimensional world. In despair that the familiar presentist view of the world contradicts relativity, some might be tempted to argue that, in addition to relativizing motion, time, space, simultaneity, the theory of relativity relativized existence as well (conveniently ignoring the fact rooted in logic and mathematics after being ultimately extracted from the physical world, which Minkowski so clearly explained – that the very possibility of relative quantities implies an underlying absolute reality!). If existence were relativized, for each of two observers A and B in relative motion only his space would exist (and nothing more) and the three-dimensionality of the world would be saved at the price of relativizing the world's existence. Such an argument does reveal utter despair because relativized existence is very close to the nonsense "nonexistent existence." For example, for each of the observers A and B only his space (and therefore his three-dimensional world, i.e., his present) would exist, whereas the space of the other observer would be nonexistent existence – A knows (due to the theory of relativity) that B has a different space which therefore exists for B, but does not exist for A, and vice versa. Kurt Gödel found it necessary to comment on the possibility to relativize existence with a single sentence [16]: "The concept of existence [. . .] cannot be relativized without destroying its meaning completely." And this is not just the intuition of the famous logician (who also has contributions to general relativity) but certainly summarizes the result of a rigorous logical and philosophical analysis of what relativized existence may mean. All who tend to suspect that Gödel might have made an unfounded statement could try to disprove it by showing that existence can be relativized without destroying its meaning.

The proof of the reality of Minkowski's absolute four-dimensional world by the theory of relativity would not be the greatest and most spectacular proof in the history of science if it relied to the slightest extent on some authorities in science; the only authority on which this proof relies is the ultimate judge – the experimental evidence. Nature has been good to us since she allowed for even such an abstract logical and philosophical issue – relativization of existence – to be tested experimentally. In Chapter 5, where we will see how such a fundamental issue as the dimensionality of the world can be tested by experiment, we will also see that the experiments, which confirmed the twin 'paradox' effect, rule out the relativization of existence (those experiments would be impossible if existence were relativized). And not only this – Chapter 5 contains an additional and independent argument, also based on the relativistic experimental evidence, which demonstrates what Gödel stressed – that relativized existence leads to nonsensical results and situations.

To summarize, even in this introductory chapter we identified a number of illusions and implicit assumptions regarding the interpretation of our sense data about the world, which have been taken for granted. When these assumptions are made explicit it becomes evident that our sense data allow for another interpretation, which is counter-intuitive. We saw (and will see in greater detail in Chapter 5) that the theory of relativity and the relativistic experimental evidence not just supported, but *proved* the counter-intuitive interpretation of the sense data.

In view of this, we can also summarize what we should trust to avoid believing in illusions:[3]

- We should trust and share a world view that is based on reliable scientific knowledge. In this way we can more easily understand and include in our world view even worryingly counter-intuitive features of the world, if the experimental evidence tells us that this is what the world is like, no matter whether we like it or not.

- We should trust that science does provide reliable (and therefore *eternal*) knowledge about the world since a scientific theory, whose predictions have been confirmed by experiment in its domain of applicability, will never be disproved in that domain by a more modern theory. The need for a thorough understanding of this issue arises from an occasional temptation that we should not worry about the implications of a scientific theory for our world view since sooner or later it would be replaced by a more modern theory which might not have such implications. As we saw above a very important lesson from the history of science is that no new theory can challenge the reliable knowledge about the world in the domain of applicability of a scientific theory where it has been proved by experiment.

The above discussion of these guiding principles is mostly intended for a wider audience without a strong scientific background. As the presentation of the arguments for the reality of Minkowski's four-dimensional world was also meant for this audience, I would like to emphasize that scientists and particularly physicists are well-aware of Minkowski's introduction of the absolute four-dimensional world (and his arguments for it) when he gave the spacetime formulation of Einstein's special relativity. There have been many physicists who have demonstrated in writing their brilliant understanding of the impact of Minkowski's idea on our view of the world.[4] Here are several examples.

A. Einstein, *Relativity: The Special and General Theory* (Routledge, London 2001) p. 152:

---

[3] I think readers of this book may find Matthew Hutson's *The 7 Laws of Magical Thinking: How Irrational Beliefs Keep Us Happy, Healthy, and Sane* [17] an excellent and helpful companion.

[4] An excellent example of how philosophers adopt the implications of relativity and Minkowski's views is G. Nerlich's book *Einstein's Genie: Spacetime out of the Bottle* [18], which was just published. I think this is a major work on the metaphysics of spacetime.

It appears therefore more natural to think of physical reality as a four-dimensional existence, instead of, as hitherto, the *evolution* of a three-dimensional existence.

A. Einstein in his letter of condolences to the widow of his longtime friend Michele Besso (Besso left this world on 15 March 1955; Einstein followed him on 18 April 1955) [15]:

Now Besso has departed from this strange world a little ahead of me. That means nothing. People like us, who believe in physics, know that the distinction between past, present and future is only a stubbornly persistent illusion.

A. S. Eddington, *Space, Time and Gravitation: An Outline of the General Relativity Theory* (Cambridge University Press, Cambridge 1920), p. 51:

In a perfectly determinate scheme the past and future may be regarded as lying mapped out – as much available to present exploration as the distant parts of space. Events do not happen; they are just there, and we come across them.

A. S. Eddington, *Space, Time and Gravitation: An Outline of the General Relativity Theory* (Cambridge University Press, Cambridge 1920), p. 56:

However successful the theory of a four-dimensional world may be, it is difficult to ignore a voice inside us which whispers: "At the back of your mind, you know that a fourth dimension is all nonsense." I fancy that that voice must often have had a busy time in the past history of physics. What nonsense to say that this solid table on which I am writing is a collection of electrons moving with prodigious speeds in empty spaces, which relatively to electronic dimensions are as wide as the spaces between the planets in the solar system! What nonsense to say that the thin air is trying to crush my body with a load of 14 lbs to the square inch! What nonsense that the star cluster which I see through the telescope obviously there now, is a glimpse into a past age 50 000 years ago! Let us not be beguiled by this voice. It is discredited.

In the distant 1921 Eddington made his most explicit comment on the reality of spacetime (Minkowski's four-dimensional world) when he discussed the fact (discovered by Minkowski as mentioned above) that not only do observers in relative motion have different times but they also have different spaces, which however are fictitious since according to the theory of relativity the world is not objectively divided into such spaces and times (A.S. Eddington, The Relativity of Time, *Nature* **106** (1921) pp. 802–804, p. 803):

> It was shown by Minkowski that all these fictitious spaces and times can be united in a single continuum of four dimensions. The question is often raised whether this four-dimensional space-time is real, or merely a mathematical construction; perhaps it is sufficient to reply that it can at any rate not be less real than the fictitious space and time which it supplants.

H. Weyl, *Philosophy of Mathematics and Natural Science* (Princeton University Press, Princeton 2009) p. 116:

> The objective world simply *is*, it does not *happen*. Only to the gaze of my consciousness, crawling upward along the life line of my body, does a section of this world come to life as a fleeting image in space which continuously changes in time.

H. Weyl, *Mind and Nature: Selected Writings on Philosophy, Mathematics, and Physics* (Princeton University Press, Princeton 2009) p. 135:

> The objective world merely exists, it does not happen; as a whole it has no history. Only before the eye of the consciousness climbing up in the world line of my body, a section of this world "comes to life" and moves past it as a spatial image engaged in temporal transformation.

R. Geroch, *General Relativity: 1972 Lecture Notes* (Minkowski Institute Press, Montreal 2013), p. 7:

> There is no dynamics in spacetime: nothing ever happens there. Spacetime is an unchanging, once-and-for-all picture encompassing past, present, and future.

However, there are scientists (even physicists) who most probably regarded the claim that the world is four-dimensional as obviously wrong due to its hugely counter-intuitive implications. I am inclined to think that the reason for such ignoring of the arguments for the reality of spacetime is rather irrational. Such an attitude toward arguments for worryingly counter-intuitive new discoveries was exhibited by Cantor in a letter to Dedekind in 1877 where he explained how he viewed one of his own major results (the one-to-one correspondence of the points on a segment of a line with (i) the points on an indefinitely long line, (ii) the points on a plane, and (iii) the points on any multidimensional mathematical space) – "I see it, but I don't believe it" [19]. This book will repeatedly stress what guides scientists in their quest for understanding the world – that the nature of the world (no matter how counter-intuitive it may be) is ultimately revealed by the experimental evidence.

Anyone who disagrees with Minkowski's view of the world, should try to avoid presenting arguments against it based on other experiments or theories. First, experiments do not contradict one another. Second, as we know science does not work in this way – if we have an argument we face it, we do not ignore it and bring other arguments. The only scientific approach would be to disprove the arguments that the relativistic experiments proved that the world is four-dimensional, i.e., anyone who tries to disprove the spacetime view should disprove the arguments showing that both the theory of relativity and the experiments which confirmed its predictions are impossible in a three-dimensional world. As we saw above and as we will see in Chapter 5, those arguments can be fully understood by non-experts and therefore non-experts are also in a position to try to refute them; the only thing they have to trust is the information about the experiments that confirmed the relativistic predictions, but that information can be verified easily.

Let me assure you that I am perfectly aware of how difficult it is to accept and especially to adopt Minkowski's totally counter-intuitive view of the world. When I first realized its huge implications for virtually all areas of our life (this happened years ago in graduate school after an advanced course in electrodynamics and during a course in general relativity), my reaction was perhaps similar to the reaction of a lot of you – the world could not be that idiotic (I am sorry for this expression, but I was very emotional when I realized how utterly absurd a four-dimensional world view looked like and this was exactly my reaction). However, instead of throwing out all my books on relativity and hoping that my refusal to accept that view would make it wrong,

I chose to follow the path of the scientific method. As like anyone else in the scientific field, I also recognize the experimental evidence as the ultimate judge and the only authority in science, I started to analyze the experiments which confirmed the relativistic effects with the firm intention to disprove Minkowski's view (i.e., the spacetime view of the world). But the analyses did not produce the results I was sure they would produce. Quite the opposite – it turned out that those experiments would be impossible if Minkowski's view were wrong, i.e., if the world were three-dimensional. After repeating those analyses, finally I asked myself – If the world is indeed a four-dimensional block ('frozen') universe, what should I do? Deny Minkowski's view (which is proved by the experimental evidence) simply because I do not like it? Now you know my answer – it is in front of you. And gradually I realized that the spacetime view is not as troubling as it might look at first sight.

I would not be completely surprised if there are people who might continue to believe in presentism without even trying to understand the arguments against it (which could unable them to change their world view). My hope is, if there are such people who are not amenable to outside help, that they will be able to help themselves for a very simple reason. No one can impose on us a given (even scientific) view – we are all entitled to our own views. However, we all know that in such situations there is always a small problem – Nature does not care about our personal opinions.

A century after Minkowski we all should finally face the facts which show that what appears to be self-evident to us – that the world exists only at the moment 'now' – is, as Einstein put it, "only a stubbornly persistent illusion." It is true that the view of reality which is consistent with modern science poses great challenges of its own. But taking refuge from the blinding light of truth back into the deceivingly safe and comfortable cave of ignorance should not be an option for anyone in the 21st century.

2 Logical Arguments Against Illusions – Eleatics, Aristotle, and Augustine

Twenty five centuries ago the quest for understanding Nature in the Western civilization had begun. The ancient Greek thinkers had started to ask questions about the nature of reality and to make the first contributions to its understanding. One of those contributions is the discovery that our senses do not provide us with unambiguous information about the world, which could give rise to illusions.

Raphael's *School of Athens* (1505)

As we saw in the previous chapter, by using the allegory of the cave Plato hinted that many things about the world, which we regard as self-evident, may in fact be illusions. This is not surprising since he had been strongly influenced by the ideas of the Eleatic school of thought, according to which the true world is dramatically different from the illusory image of it provided by our senses. The Eleatic school, whose representatives were Parmenides, Zeno, Melissus, and perhaps Xenophanes, is a unique case in the intellectual history of our civilization. Twenty-five centuries ago the Eleatics had more trust in reason than in what appears to follow from our senses.

The essence of the Eleatic view had been developed by Parmenides in his poem *Peri Physeos*. It presents two paths of knowledge – one leading to the truth (the way of Truth), the other to the opinions of men (the way of Opinion). The second path of knowledge deals with the world of our perceptions. This shows that the Eleatics did not deny what we perceive, but held that it is rather illusory and is therefore not revealing the ultimate reality.

Parmenides (born ca. 510 BC)

However, the Eleatics denied what had been taken as self-evident by Heraclitus and other thinkers – that what exists, exists only at the present moment. Such an understanding of our everyday experience, according to the Eleatics, meant that what was in the future and did not exist, comes into existence by becoming present, and what existed as present goes out of existence by becoming past. Parmenides believed that nothing could come into or go out of being (existence) because it would contradict a basic principle – *being exists, non-being does not exist*. Parmenides held that this principle (which might be also viewed as an ancient form of the fundamental idea of conservation – something cannot arise out of nothing) could be deduced from what we all perceive [20, p. 64-66]:

> There are very many signs: that Being is ungenerated and imperishable, entire, unique, unmoved and perfect; it never was nor will be, since it is now all together, one, indivisible.

The immediate implications of this basic principle are both profound and completely counter-intuitive:

- Being is eternal (ungenerated) and unchanging – it does not arise and does not disappear.

- Being is indivisible – continuous and homogeneous.

Indeed the Eleatics argued that their basic principle (deduced through a rigorous analysis from what we all observe) reveals the true reality as an *eternal* entity – an eternal present since nothing can come into being and become present, and nothing can go out of being, because everything that exists, exists as this eternal present and nothing else exists. If something were to come into being it should come either from being or from non-being, but neither of these alternatives is possible since being cannot become being (it is already being), and being cannot come from non-being either since non-being does not exist. By the same argument, nothing can go out of existence – if something were to go out of being it should transform either to being or to non-being; but being cannot become being (it is already being), and being cannot become non-being because such a thing does not and cannot exist. Here is probably the original form of this argument as stated by Gorgias, another pre-Socratic philosopher (quoted from [21, p. 28]):

> What is cannot have come into being. If it did, it came either from what is or what is not. But it did not come from what is, since if it is existent it did not come to be but already is; nor from what is not, for the non-existent cannot generate anything.

The Eleatics regarded the flow of time (the perpetual transformation of the non-existent future into the existent present, and the existent present into the non-existent past) as self-contradictory and argued that time does not belong to the true reality and is merely an illusion [20, pp. 74-78]:

> And time is not nor will be another thing alongside Being, since this was bound fast by fate to be entire and changeless. Therefore all those things will be a name, which mortals, confident that they are real, suppose to be coming to be and perishing, to be and not to be . . .

The second implication of the Eleatics' basic principle is no less counter-intuitive. How could being be indivisible (continuous and homogeneous) given the fact that our daily experience tells us the opposite – that being is anything but homogeneous. The Eleatics did not deny that there is nothing homogeneous in what we see every day, but argued that what we see is a continuous illusion.

Now, in the beginning of the 21st century many might try to explain the Eleatics' insistence that being is homogeneous, and that the huge variety of different things around us is indeed rather illusory, by referring to modern physics – that the huge variety of things is a manifestation of only a small number of building blocks (elementary particles) and ultimately a manifestation of quantum fields. Such an explanation is in a sense richer than that meant by the Eleatics who had no knowledge of the Standard Model and quantum field theory. By arguing that being is indivisible (homogeneous) they had in mind something simple but profound – the most fundamental feature of all different things, which we see, is their existence; in this sense we see utter homogeneity – only existent things.

Now it is clear why the Eleatics regard being as indivisible. If we assume that being is divisible (inhomogeneous) we should explain what is dividing it. Being is the only thing that exists and therefore it cannot be divided by something else since nothing exists except being. Formally, there are two possibilities – being is divided either by being or non-being. But obviously being cannot divide being since what divides is the same thing that should be divided and therefore no division is possible; we cannot say water is divided by water. Non-being cannot divide being either since something non-existent cannot divide anything.

Parmenides appeared to have been fully aware of how the Eleatic view of the true reality would be received by the other people. That is why in his poem he examined the second path of knowledge which deals with the world of our perceptions. The purpose of this part of the poem is nicely explained by Guthrie [21, pp. 5-6]:

> Suppose that Parmenides is doing his best for the sensible world, perhaps on practical grounds, by giving as coherent an account of it as he can, saying in effect: I have told you the truth, so that if you go on to speak about the world in which we apparently live you will know it is unreal and not be taken in. But after all, this is how it does appear to us; however misleading our senses may be, we must eat and drink and talk, avoid putting our hand in the fire or falling over a precipice, live in short as if their information were genuine. Being ourselves mortals we must come to terms with this deceitful show, and I can at least help you to understand it better than other people.

Despite Parmenides' efforts, naturally, the ideas of the Eleatics have been always viewed as extreme. As no one has succeeded in coming even close to disproving them the most often reaction was to ridicule them (since it is not necessary to understand something in order to ridicule it). Perhaps the most serious criticism of the Eleatics' world view was the attack against Parmenides' principle – only being exists, while non-being does not exist. The argument against it was that it is an obvious tautology – what exists exists, what does not exist does not exist. However, Jonathan Barnes, a prominent pre-Socratic scholar, pointed out that Parmenides in fact "maintains that whatever exists exists and cannot not exist" and for this reason his central assertion "does not turn out tautologous; since it is far from a tautology that what exists *cannot not exist*" [22].

That this single (deserving attention) argument against Parmenides' principle does not harm it, is best seen if we accept the argument – that, formally, Parmenides' principle as defined in this chapter (which is perhaps the most common translation) appears to be a tautology. Now recall that a definition of something is in terms of something more fundamental. Then it becomes clear that being – the most fundamental concept – cannot be defined in terms of something more fundamental. Therefore, the only possible definition of the most fundamental concept is unavoidably tautological – being is defined through itself or its negation – non-being.

Zeno (ca. 490 BC – ca. 430 BC) shows the Doors to Truth and Falsity

Zeno's role in the Eleatic school was not so much to develop Parmenides' arguments, but to defend the Eleatic view indirectly. Zeno "reversed the logic" and told those who ridiculed the Eleatic thinkers – you say Parmenides' view that being is one eternal and homogeneous entity is absurd, but I will show you that your own view that the illusory world you perceive is the true reality leads to absurd consequences. This is clearly seen when in Plato's *Parmenides* Zeno explains to Socrates the purpose of his book [23, p. 362]:

> The truth is that the book comes to the defense of Parmenides' argument against those who try to make fun of it by claiming that, if it is one, many absurdities and self-contradictions result from that argument. Accordingly, my book speaks against those who assert the many and pays

them back in kind with something for good measure, since
it aims to make clear that their hypothesis, if it is many,
would, if someone examined the matter thoroughly, suffer
consequences even more absurd than those suffered by the
hypothesis of its being one.

The true reality according to the Eleatic view is a changeless being,
whereas our daily experience involves mostly observations of different
kinds of motion and changes. It is those observations that have been
offered as proof that the Eleatic view was an obvious absurdity. For
this reason, Zeno formulated a number of paradoxes intended to show
that if one assumed that motion were real, a careful analysis would
show that motion was in fact impossible. In Part VI of his *Physics*
Aristotle wrote [24, §9, 239b10-239b13]:

Zeno's arguments about motion, which cause so much trou-
ble to those who try to answer them, are four in number.
The first asserts the non-existence of motion on the ground
that that which is in locomotion must arrive at the half-
way stage before it arrives at the goal.

Let us examine the first argument (paradox) by Zeno, which is
called the *Dichotomy* since it involves repeated division into two as
explained by Aristotle. Ironically, although Zeno's argument failed
(which had no effect on the Eleatic arguments since Zeno's argument
is not based on them), its failure further boosted the Eleatic view
by setting the stage for a powerful logical argument against the sole
existence of the present moment. This argument demonstrates that
one does not need to agree with the Eleatic view to realize that not
only does the presentist view (discussed in the previous chapter) fail
to explain how something can come into and go out of existence, but
also that such a view (based on the assumption that only the present
moment is real) appears to be self-contradictory.

It seems Aristotle was the first who realized that self-contradiction
when he resolved Zeno's paradox *Dichotomy*. Zeno argued that an
object moving from a point A to a distant point B would never reach
B since it would need an infinite amount of time, first to travel half of
the distance AB, then half of the remaining half, and so on to infinity.
Aristotle showed that Zeno had arrived at the paradox, because he
explicitly presupposed that space was divisible to infinity, but implicitly assumed that time was not infinitely divisible (if both space and
time are infinitely divisible, there is no paradox – if, for example, a
distance of one meter is traveled by a body for one second, the body

will travel half a meter for half a second and so on, and will not need an infinite amount of time to reach the end point B). In Book VI of *Physics* Aristotle wrote about Zeno's implicit assumption that time is not infinitely divisible [24, §9, 239b5-239b9]:

> This is false; for time is not composed of indivisible nows any more than any other magnitude is composed of indivisibles.

Aristotle resolved Zeno's paradox by stressing that time, like space, is also infinitely divisible. Throughout Book VI he repeatedly stated that time was infinitely divisible – e.g., "all time is infinitely divisible" [24, §8, 238b37-239a10] and "all magnitudes and all periods of time are always divisible" [24, §6, 237b10-237b20].

Let us summarize Aristotle view on the divisibility of time as *Statement 1*: *time is not composed of indivisible nows.*

However, when Aristotle discussed the nature of time itself in Book IV of *Physics* – that of all times (past, present, and future) only the moment 'now' is real – his logical analysis unavoidably led him to the *opposite* conclusion: that the only real moment of time is "the indivisible present 'now' " [24, §13, 222b9-222b14]. In Book VI he explained what forced him to contradict himself and effectively to admit

Aristotle (384 BC – 322 BC)

that Zeno succeeded in proving that motion was impossible (since Zeno's paradox is based on the assumption that time is not infinitely divisible) [24, §3, 234a5-234a23]:

> All time has been shown to be divisible. Thus on this assumption the now is divisible. But if the now is divisible, there will be part of the past in the future and part of the future in the past ... It is clear, then, from what has been said that time contains something indivisible, and this is what we call the now.

Aristotle realized that he had no choice but to talk about "the indivisible present 'now' " in order to avoid a contradiction in terms – if the moment 'now,' which by definition is *wholly present*, were divisible, it would contain past, present, and future moments (e.g., if

36

the moment 'now' lasted, say, ten seconds and if it were divisible, half of it would be past and the other half future, with the present second in the middle of the interval).

We can summarize Aristotle's view on the divisibility of the moment 'now' as *Statement 2*: *time contains something indivisible, and this is what we call the now.*

Aristotle's *Statement 1* and *Statement 2* are perhaps in one of the most fruitful contradictions – the type of contradictions and paradoxes that as if want to tell us something profound about the world. As N. Bohr put it (quoted from [25]):

> How wonderful that we have met with a paradox. Now we have some hope of making progress.

The very fact that Aristotle, one of the greatest thinkers of our civilization who single-handedly created the science of logic, was led by the common-sense view that only 'now' is real to the contradiction – the present moment is both divisible and indivisible – implies that the presentist view's basic assumption of the sole existence of the moment 'now' is wrong. Aristotle seems to have tried to identify the cause of that contradiction. An indication of such an attempt in Book IV of his *Physics* is his doubt on whether the division of time into past, present, and future

Saint Augustine
13 November 354 – 28 August 430

reflected an objective fact or that division had something to do with the mind (or the soul) [24, §14, 223a22-223a28]:

> Whether if soul did not exist time would exist or not, is a question that may fairly be asked.

Sixteen centuries ago Augustine also investigated the nature of time and like Aristotle faced the same paradoxical situation about the duration of 'now,' but unlike him explicitly concluded that the division of time into past, present, and future does not reflect an objective feature of the world and therefore should belong to the mind [26, p. 235]:

> What is by now evident and clear is that neither future
> nor past exists, and it is inexact language to speak of three
> times – past, present, and future [...] In the soul there are
> these three aspects of time, and I do not see them anywhere
> else.

One might be tempted to say that Aristotle and Augustine could
have avoided the paradox with the duration of 'now' by assuming that
it is zero. Aristotle seems to have regarded this option as obviously
unacceptable and had not even bothered to discuss it. And, indeed,
on the presentist view that option is ruled out – if the duration of the
present moment were zero it follows that even 'now' would not exist
(since zero means non-existence) and therefore *no part of time would
exist*.

As the duration of the moment 'now' cannot be zero it should be
finite. But a finite 'now' should not be divisible because otherwise it
would constitute a contradiction in terms. The only remaining option
is a finite 'now,' which is indivisible. The analysis of this option (i.e., of
the duration of the present moment) reveals an *intrinsic link between
time and the dimensionality of the world*. This link is immediately
shown by the fact that a finite indivisible 'now' challenges one of the
basic assumptions in the presentist view – that physical objects and the
world are three-dimensional, which has been regarded as self-evident
since Aristotle[1].

The challenge is best comprehended by assuming, for the sake of
the argument, that 'now' lasted, say, ten seconds. This would mean
that physical objects and the world would not be three-dimensional
since they would exist *at once at all moments* of the ten-second 'now'
and would be therefore *extended in time*. As a concrete example consider the motion of a car moving at speed 60 km/h. As the ten seconds
are not further divisible into past, present, and future, the car would
exist *equally (at once) at all moments* of the ten-second 'now'. This
means that it would be spread over a distance of 166.7 meters (note
that every point of the car, e.g. the point, where the licence plate
is, would be spread over that distance of 166.7 meters) and therefore would not be a three-dimensional object – it would be a four-
dimensional object which extends ten seconds in time. This example
also demonstrates that the car would be the familiar three-dimensional

---

[1]In Book I of his *On the Heavens* Aristotle wrote [24, §1, 268a1-268b10]: "A
magnitude if divisible one way is a line, if two ways a surface, and if three a body.
Beyond these there is no other magnitude, because the three dimensions are all
that there are."

object *only* if 'now' lasted zero seconds – only then all points of the car would be *points* and not lines[2] consisting of the car's points smeared out over the distance of 166.7 meters (to see this even easier you could assume that 'now' lasted 0.1 seconds – then the point, where the licence plate is, would be spread over a distance of 1.7 meters; in the case of car travelling at 120 km/h the point, where the licence plate is, would obviously be spread over a distance of 3.4 meters). As zero duration of 'now' means that 'now' does not exist (and therefore time does not exist since on the presentist view the only part of time that is real is 'now'), it is clear that the world would be three-dimensional if time did not exist.

If the ancient logical arguments discussed above had been seriously examined it would have been realized that the presentist view is self-contradictory. Presentists have been taking for granted two things which upon a closer scrutiny turn out to be contradictory:

- only the present moment (the moment 'now') exists

- the present (the world at the moment 'now') is three-dimensional.

We just saw that these two basic assumptions exclude one another – in order for the moment 'now' to exist it should be finite (*not zero*) and indivisible, which means that the world would not be three-dimensional, but extended in time. The world would be three-dimensional *only* if the duration of 'now' were zero, which means that 'now' would not exist and therefore no part of time would exist.

If the option of a finite, but indivisible 'now' had been thoroughly analyzed before the advent of special relativity, it would have been possible to examine the nature of time more rigorously and to distinguish two aspects of time whose mixture has been causing a lot of confusion over the centuries – the dynamical one, according to which only one moment of time – the constantly changing 'now' – exists, and the geometric (or static) one, which reveals the resemblance of time to space (e.g. the resemblance of the interval between two moments of time to the distance between two points in space). Such an analysis would have strengthened the suspicion that the dynamical aspect of time is based exclusively on the fact that we realize ourselves and the world at the constantly changing present moment and uncritically

---

[2]Think of the concept of a point – a point, which is regarded as zero-dimensional, is not extended in any dimension; otherwise it will be a line, i.e., a one-dimensional object. Analogously, a surface is a two-dimensional object with zero extension (thickness) in the third dimension; if the extension of the surface in the third dimension were not zero, the surface would be a three-dimensional object.

assume that only 'now' exists and that time objectively flows. Then the question of whether the world itself also exists only at the moment 'now' could have been explicitly asked.

3 GALILEO'S BRILLIANT MIND – EXPERIMENTS AND LOGIC AGAINST ILLUSIONS

In this chapter we will discuss Galileo's contributions to our deeper understanding of the nature of reality. In addition to the crucial improvements of the telescope which allowed him to make contributions to observational astronomy, Galileo made other contributions to science so important that essentially Galileo's achievements (before the Newtonian revolution) are now regarded as marking the birth of modern science in the 17th century.

Galileo's other contributions to science can be divided into two groups. The first group consists of two general contributions which formed the foundations of the art of doing science:

Galileo Galilei
15 February 1564 – 8 January 1642

- What is now called the scientific method – it comprises three elements which constitute the essence of scientific research:

  - formulating hypotheses to *explain* observations,

  - deducing consequences (predictions) from the hypotheses,

  - using *experiments* to test those hypotheses by testing their predictions.

41

We owe the clear realization, that it is the scientific method that distinguishes science from all other human activities, mostly to Galileo.

- A powerful and productive research strategy – exploring the internal logic of ideas – which was most successfully and consistently employed by Galileo himself as well as by Einstein as we will see in Chapter 4 and Chapter 6. Exploring the internal logic of an idea can be briefly described as deducing all logical consequences of the idea, examining them through thought experiments, and testing them with real experiments.

The second group of Galileo's contributions contain his major scientific achievements:

- Disproving the twenty-century-old Aristotelian view of motion according to which any motion needed a mover (excluding celestial motion and the vertical motion of some elements, up or down, which Aristotle regarded as natural) and replacing it with the adequate view of motion – that bodies moving uniformly (with constant velocity) move on their own by inertia and therefore do not need a mover.

- Disproving the centuries-old belief that rest and motion are experimentally distinguishable states by showing that uniform motion cannot be discovered by mechanical experiments. We now call this discovery Galileo's principle of relativity.

- Disproving the centuries-old belief that heavier bodies fall faster toward the Earth's surface and demonstrating that all bodies fall equally fast.

Let us examine how Galileo identified these centuries-old illusions and how he arrived at his discoveries.

In our daily experience we encounter different kinds of motion virtually at every instant and because of our familiarity with this phenomenon we regard it as completely self-evident and self-explanatory. However, motion turns out to be one of those familiar phenomena which Nature has given us as apparently self-evident, without adequate understanding. Since ancient times thinkers have suspected that a deeper nature hides behind the motion's apparent manifestations in our daily life and have tried to understand it. As we saw in the previous chapter the Eleatics argued that motion did not belong to the true reality and for this reason was nothing more than an illusion. But even those thinkers who believed in the reality of motion struggled to understand its nature.

In ancient Greece Aristotle (384 BC – 322 BC) devoted a significant part of his *Physics* to the phenomenon of motion. He examined how bodies move and concluded that in "some cases their motion is natural, in others violent and unnatural" [27, p. 339 (Book VIII)], that is, "all movement is either compulsory or according to nature" [27, p. 294 (Book IV)]. Natural motion or motion "according to nature" was mostly defined on the basis of his view that everything in the terrestrial world consisted of four elements (unlike the heavens made of quintessence, i.e. a fifth element) – air, earth, fire, and water – and the assumption that the elements tend to move towards their natural places, e.g. "a body has a natural locomotion towards the center if it is heavy, and upwards if it is light" [27, p. 284 (Book III)]. Aristotle also defined another type of natural motion: "Thus in things that derive their motion from themselves, e.g. all animals, the motion is natural (for when an animal is in motion its motion is derived from itself): and whenever the source of the motion of a thing is in the thing itself we say that the motion of that thing is natural" [27, p. 339 (Book VIII)].

On Aristotle's view natural motion is harmonious, not forced motion. A body which moves naturally, moves on its own; it is not compelled to move by something external. For example a body falling towards the surface of the Earth moves naturally, since the basic element comprising the body – earth – tends to occupy its natural place – the center of the Earth. Natural motions on Earth (excluding animals) are downwards and upwards (air tends to go up), whereas the natural motion of celestial bodies is in circles.

According to Aristotle, unnatural (violent) motion is disharmonious since it is a forced motion. That is, any object (again excluding animals) which moves not vertically is prevented from moving naturally and therefore must be moved by something. Aristotle stressed that in the case of unnatural (violent) motion "Everything that is in motion must be moved by something" [27, p. 326 (Book VII)].

Aristotle's view of motion and his geocentric model of the Universe with the Earth at the center (further developed by Ptolemy (90 – ca. 168)) dominated the pre-scientific understanding of the world for twenty centuries. The geocentric, or Ptolemaic, system had been universally accepted throughout those centuries despite that shortly after Aristotle another Greek – Aristarchus of Samos (310 BC – ca. 230 BC) – developed the first heliocentric model of our planetary system placing the Sun at the center [28]. His model was in contrast to another alternative to the geocentric model, which was proposed before Aristotle by the Pythagorean Philolaus (ca. 470 BC – ca. 385 BC), in which all celestial bodies, including the Sun, revolved around a central fire.

Although Aristarchus himself accurately answered (as we now know) the only relevant astronomical argument against his model – that there should be observable stellar parallax if the Earth orbited the Sun – by assuming that the stars are very far away, the heliocentric model of the solar system did not survive. There have been many attempts to explain why the correct explanation had failed and some believe that a kind of intellectual conservatism might have been responsible for that [28, p. 39]:

> There are many theories, all very speculative, of why the heliocentric theory did not catch on and was superseded so completely by geocentric astronomy. Perhaps the most plausible is the simple fact that the geometrical skill of the Greeks allowed them to devise ingenious constructions that modelled accurately the motions of all the heavenly bodies without having to take the drastic step of removing the Earth from its privileged position at the centre of the Universe.

I think the very way the geocentric model was refuted, as we will now see, demonstrates that the most probable reason is that the heliocentric model contradicted Aristotle's view of motion. Here are some of the early examples of that contradiction summarized by Ptolemy in his *The Almagest* in which he implicitly used Aristotle's view of motion to show that if the Earth were not stationary but rotated around its axis we would observe some effects of that motion, but such effects were never detected (the view that the Earth orbits the Sun requires that it also turns about its axis to explain the change of day and night) [29]:

> They would have to admit that the earth's turning is the swiftest of absolutely all the movements about it because of its making so great a revolution in a short time, so that all those things that were not at rest on the earth would seem to have a movement contrary to it, and never would a cloud be seen to move toward the east nor anything else that flew or was thrown into the air. For the earth would always outstrip them in its eastward motion, so that all other bodies would seem to be left behind and to move towards the west.

These arguments against the view of a moving Earth remained unchallenged until the seventeenth century when Galileo (1564 – 1642) succeeded in refuting them by first refuting Aristotle's view of motion. The fact that the second and more detailed heliocentric model of the solar system, proposed by Copernicus (1473 – 1543) before Galileo's birth, was accepted only after Galileo's refutation of Aristotle's doctrine of motion, does seem to indicate that Aristarchus' model ultimately failed because it contradicted that doctrine.

Here is Galileo's first lesson of how exploring the internal logic of an idea behind an illusion can lead to exposing the illusion. In the second chapter (The Second Day) of his book *Dialogue Concerning the Two Chief World Systems – Ptolemaic and Copernican* Galileo discussed mostly a single experiment – the tower experiment – which summarized the arguments which had been regarded as proving the motionlessness of the Earth. The tower argument can be formulated in the following way: if the Earth were not

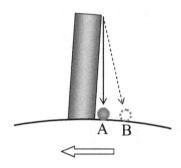

According to the supporters of the Ptolemaic system of the world, if the Earth were revolving around its axis and a stone is dropped from the top of, say, the Leaning Tower in Pisa it would hit the ground at B, not at A.

stationary but were turning around its axis, then a stone dropped from the top of a tower would not fall at the base of the tower (point A), but at point B, since during the time taken by the stone to fall, the tower (being carried by the Earth's turning) would travel the distance AB. Galileo clearly realized that it was Aristotle's view of motion – everything that is in motion must be moved by something – which was behind this argument. Indeed, according to the Aristotelian view, when the stone is held on the top of the tower it has a mover which moves it horizontally together with the tower (and is prevented from following its natural tendency to fall toward the center of the Earth). However, when the stone is released it only moves naturally (falls), but there is no mover to force it to move also horizontally and the stone is left behind the tower and hits the ground at point B.

It is quite possible that Galileo himself dropped stones from the top of the mast of both a moving ship (or observed such experiments) and saw that the stone always fell at the foot of the must [30, pp. 144-145]:

For anyone who does will find that the experiment shows

exactly the opposite of what is written; that is, it will show
that the stone always falls in the same place on the ship,
whether the ship is standing still or moving with any speed
you please. Therefore the same cause holding good on the
earth as on the ship, nothing can be inferred about earth's
motion or rest from the stone falling always perpendicu-
larly to the foot of the tower.

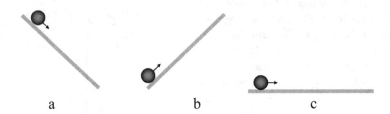

a   b   c

A simple analysis of the motion of a ball on inclined and horizontal planes led Galileo
to the idea of inertial motion. Such an analysis could have been carried out by
Aristotle and by all thinkers who for twenty centuries had supported his view of
motion. Galileo did not possess any piece of extra knowledge that was not available to
his predecessors.

Instead of stating clearly that he did perform or observe the experiments on a moving ship, Galileo chose to demonstrate the power of
his method of exploring the internal logic of ideas to arrive at the same
result: "Without experiment, I am sure that the effect will happen as
I tell you, because it must happen that way" [30, p. 145].

He did that by first refuting Aristotle's view of motion since it is
this view that is behind the belief that rest and motion can be detected
experimentally. Galileo demonstrated that Aristotle's doctrine of motion was wrong in the case of the motion of a firm ball on "a plane
surface as smooth as a mirror and made of some hard material like
steel" [30, p. 145] (He analyzed several similar thought experiments
apparently based on real observations). Galileo considered three cases
of the motion of a ball – on a downward slope (case *a* in the figure),
upward slope (case *b*), and on a horizontal plane (case *c*). In the
first case the ball *increases* its velocity when released, whereas in the
second case its velocity *decreases* after it was pushed upwards on the
plane.

For twenty centuries the supporters of Aristotle's view of motion
(and, of course, Aristotle himself) have been well aware of plenty of
instances of such trivial observations and experiments, but did not
bother to perform even a simple analysis of what they have been observing (which is inexplicable in the case of such a great thinker as
Aristotle whose study of the nature of motion occupies a significant

part of his *Physics*). Had Aristotle asked himself how a ball would move on a smooth horizontal surface (after being initially put in motion by a mover which is then removed), he undoubtedly would have arrived at the conclusion first realized by Galileo – the horizontal plane has neither downward nor upward slope, which means that the ball's velocity will *neither increase nor decrease*; therefore the ball will move with *constant* velocity. If the plane is infinite and the negligible friction is ignored the ball will move *forever* – "if such a space were unbounded, the motion on it would likewise be boundless? That is, perpetual?" [30, p. 147] (the friction only slows down and eventually stops the ball's eternal motion, but obviously has nothing to do with the newly discovered nature of motion since the ball moves on its own as Galileo's analysis revealed, and it is clearly not the friction that moves the ball).

In a spectacular way, by exploring the internal logic of amazingly simple everyday observations of motion, Galileo disproved Aristotle's doctrine of motion (inexplicably unchallenged for twenty centuries before Galileo) and arrived at a new view of motion – a free body moves uniformly *on its own without a mover*[1].

The analyses of such experiments marked the end of Aristotle's view that a moving body requires a mover. According to the new view of motion, extracted through analyses from the observation of mundane physical phenomena, a body moves naturally (or by inertia) when it moves uniformly (with constant velocity) on its own (without a mover), whereas forced (violent) motion occurs when a body is prevented from moving naturally.

Galileo's decision not to disclose whether he did or observed experiments of dropping a stone from the top of the mast of a moving ship might have been intended to convey a message to the future generations – whenever possible, try to explore the internal logic of ideas involved in open questions in order to arrive at resolutions even before performing real experiments. And he did give a convincing example – by employing *the new concept of inertial motion* to the *thought* experiment of dropping a stone from the top of a tower or the mast of a moving ship he proved that the stone must fall at the foot of the tower or the mast, exactly at the place where it would fall if the earth or

---

[1]Galileo's explanation of the experiments with inclined and horizontal planes unambiguously demonstrates (see specifically p. 147 of [30]) that he disproved Aristotle's view of motion and arrived at the *correct* understanding of what we now call inertial motion. That is why I completely disagree with a relatively recent suggestion that Galileo's physics (particularly his understanding of inertia) is closer to Aristotle's physics than to Newton's [31]. It is true that Newton further developed the concept of inertia distinguishing its two aspects, but Newton himself repeatedly referred to Galileo's contribution to the new view of inertial motion.

the ship were motionless: when the stone is released it preserves the horizontal component of its velocity (like the ball on the horizontal plane preserves its velocity after its mover is removed) and will fall at the base of the tower or the mast.

Therefore, as Galileo stated, it really does not matter whether or not a ship is at rest or moving; a stone dropped from the top of its mast will always fall at the base of the mast. No doubt, this had been an impressive discovery at the time when being at rest and being in motion had been regarded as two distinct states. Galileo's analysis of thought (and perhaps real) experiments unequivocally demonstrated that such experiments involving inertial motion could not distinguish rest and motion.

By analyzing real and thought experiments Galileo refuted both Aristotle's view of motion and the arguments against the heliocentric model of the solar system, and arrived at two profound results:

- the idea of inertial motion – a free body moves with constant velocity on its own[2]

- what we now call Galileo's principle of relativity – by performing mechanical experiments it is impossible to determine whether a body is at rest or moves with constant velocity.

I will complete the brief summary of these two great discoveries by Galileo with another brilliant example of his method of exploring the internal logic of ideas. The reason is to emphasize what was only stated in the beginning of this chapter – that not only did Galileo make contributions to physics, but he also gave us examples of doing physics successfully by showing how a productive research strategy (based on exploring the internal logic of ideas) works.

Galileo did not simply disprove Aristotle's view of motion (that a moving body should be *moved by something*) by demonstrating that free bodies move with constant velocity *on their own* (by inertia); he completely destroyed it as if as a revenge that for twenty centuries Aristotle's doctrine unjustifiably held up the advancement of our understanding of the world. "Unjustifiably" because Aristotle himself and

---

[2]To appreciate more fully how much we owe to Galileo's brilliant mind for the proper understanding of the nature of motion, consider a car moving at constant *velocity* (constant speed and constant direction) on the highway. Even in the beginning of the 21st century we tend to think that it is the engine that moves the car, whereas the car moves on its own (by inertia), and the engine only overcomes the resistance between the tires and the ground and the air resistance. That the car moves on its own is also demonstrated when an obstacle tries to prevent it from moving uniformly; then the car resists the change in its motion.

his followers during those centuries could have *realistically* discovered that that was a wrong view. You can judge whether this is really so from this example.

Aristotle appeared to have had a chance to disprove his own view when he had been struggling to explain the motion of projectiles. When a stone is thrown, no mover is moving it. For some reason Aristotle gave a hasty explanation – the thrower's arm moves the air when the stone is thrown and the air acts as a mover of the stone. Galileo explored the internal logic of Aristotle's idea that projectiles are moved by the air. He started with a thought experiment in which a piece of cotton and a stone are placed on a table in an open field in a windy day. It is obvious that even not a strong wind will easily move the piece of cotton but not the stone. As the wind is motion of the air Galileo argued that the thought experiment unambiguously demonstrated that air moves light object more easily than heavy objects. This disproved Aristotle's explanation because if the air were the mover of projectiles than a thrown piece of cotton would go much farther than a thrown stone.

Galileo's analysis involved everyday phenomena, which had been repeatedly observed not only by Aristotle but also by ordinary people long before him. In other words, *Galileo did not use any piece of knowledge that was unavailable twenty centuries earlier*. For this reason I think Galileo's discoveries (inertial motion and Galileo's relativity principle) are delayed discoveries whose delay is difficult to comprehend and whose implications for the delayed advancement of our civilization are impossible to estimate. The only thing that could provide some explanation of how such a great thinker as Aristotle, failed to see the problems with his view of motion, is that Galileo was the first who discovered and systematically used an innovative research strategy which turned out to be extraordinarily productive – conceptual analyses of real and thought experiments through exploring the internal logic of the ideas involved in those experiments.

Let us now see how Galileo arrived at his third major scientific achievement – that bodies of different weight fall toward the Earth's surface at the same rate. Most probably he first analyzed thought experiments and then tested his conclusions by performing real experiments.

Galileo again employed his method of exploring the internal logic to the centuries-old belief that heavy bodies fall faster than light bodies. It turned out that that belief was an illusion, which could have been identified centuries earlier because in his analysis Galileo did not use anything that was not known to Aristotle (who had believed in that illusion) and to the people before him. Here is how Galileo explored the internal logic of the idea that heavy bodies fall faster that light ones. He considered the thought experiment of tying together a heavy and a light body and dropping them from a height. Trying to figure out how the system of two bodies will fall on the basis of the idea that heavy leads to a contradiction. First, as the system is heavier than the heavy body, the system should fall faster than the heavy body. Second, as the light body falls slower than the heavy body, the system of both bodies should fall slower that the heavy body because the light body slows down the fall of the system. The obvious resolution of this contradiction is that the initial assumption (the idea that heavy bodies fall faster than light ones) is wrong. Therefore all bodies fall at equal rate.

Here is Galileo's own demonstration of how the assumption that a heavy body falls faster than a light body leads to a contradiction [30, p. 446]:

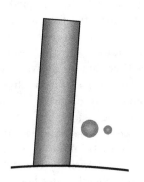

Two balls of different weight dropped from the Leaning Tower of Pisa fall at the same rate.

> If then we take two bodies whose natural speeds are different, it is clear that on uniting the two, the more rapid one will be partly retarded by the slower, and the slower will be somewhat hastened by the swifter... But if this is true, and if a large stone moves with a speed of, say, eight while a smaller moves with a speed of four, then when they are united, the system will move with a speed less than eight; but the two stones when tied together make a stone larger than that which before moved with a speed of eight. Hence the heavier body moves with less speed than the lighter; an effect which is contrary to your supposition. Thus you see how, from your assumption that the heavier body moves more rapidly than the lighter one, I infer that the heavier body moves more slowly.

Although Galileo's brilliant analysis leaves no doubt that a real experiment will confirm the conclusion of the thought experiment, it is almost certain that he did drop bodies of different weight from the Leaning Tower of Pisa and found that it takes them *same* time for the bodies to reach the ground. There have been some doubts (probably prompted by Koyré [32]) that Galileo did such experiments but Galileo's texts themselves leave no such doubts. Here is an example which implies that he did perform experiments with heavy and light bodies [30, pp. 447-448]:

> Aristotle says that "an iron ball of one hundred pounds falling from a height of one hundred cubits reaches the ground before a one-pound ball has fallen a single cubit." I say that they arrive at the same time. You find, *on making the experiment* [italics added], that the larger outstrips the smaller by two finger-breadths, that is, when the larger has reached the ground, the other is short of it by two finger-breadths; now you would not hide behind these two fingers the ninety-nine cubits of Aristotle, nor would you mention my small error and at the same time pass over in silence his very large one.

Summarizing Galileo's scientific achievements, we clearly see that we owe the identification of three particularly stubborn illusions to his brilliant mind:

- For twenty centuries people held Aristotle's view of motion – that any motion (excluding vertical or natural motion) needed a mover. Galileo explored the internal logic of that idea by using everyday observations and simple logic (all available to everyone before him, including Aristotle) and showed that that view of motion was an illusion. Galileo showed that uniform motion (motion with constant velocity) needed no mover – we now call it motion by inertia.

- For over twenty centuries people believed in the embarrassing illusion that the Earth was at the center of the world and therefore was motionless. The main reason for that view appears to have been the belief that if the Earth moved we would be able to discover its motion by many even daily observations. By employing the correct view of motion – inertial motion – Galileo demonstrated that no experiment can distinguish rest from uniform motion and therefore if the Earth moved, its uniform motion could not be detected experimentally. We now call this discovery Galileo's principle of relativity – uniform motion cannot be detected by mechanical experiments. It was this discovery that paved the way for the acceptance of the Copernican model of the solar system since Copernicus did not disprove Aristotle's doctrine of motion and therefore did not refute the arguments against the view that the Earth is moving.

- For over twenty centuries people believed in another illusion – that heavy bodies fall toward the Earth's surface faster than light bodies. By exploring the internal logic of this idea Galileo showed that it was self-contradictory. His conclusion (which he tested experimentally) was what is now a well-established experimental fact – that all bodies (no matter their weight) fall equally toward the Earth's surface (only the air resistance slows down differently the fall of different bodies).

We saw that these achievements are in fact delayed discoveries, which could have *actually* happened centuries earlier, because Galileo did not use any knowledge that was unavailable to his predecessors. It turns out that Galileo's principle of relativity *logically contains* the special theory of relativity in its four-dimensional (spacetime) formulation given by Minkowski (hardly) in 1908 [33, Chap. 3]. In this sense special relativity also appears to be a delayed discovery. However, it is of different type since it does not appear realistic to think that special relativity and the idea of spacetime could have been actually discovered soon after Galileo arrived at his principle of relativity; simply, the idea of spacetime (and therefore of special relativity as a theory of the physics of spacetime) is logically contained in Galileo's principle of relativity as we will now see.

Both the actual discovery of special relativity by Einstein in 1905 and its spacetime formulation by Minkowski in 1908 followed a path that essentially relied on Maxwell's electrodynamics, which means that special relativity could not have been discovered before Maxwell's discovery of his equations. But it could have been realistically discovered at least twenty years earlier, i.e. before the Michelson-Morley experiments since Einstein himself admitted that "Michelson's experiment played no role" [34, p.172] in the discovery of his special relativity.

Here I will summarize the results of a detailed analysis of Galileo's principle of relativity given in [33, Chap. 3] for two reasons. First, to demonstrate the power of Galileo's research method of exploring the internal logic of ideas when applied to his principle of relativity. Second, this principle has profound implications for the nature of reality since the idea of spacetime is hidden deep in that principle. This path to the discovery of the concept of spacetime is different from the actual discovery of special relativity and its spacetime formulation, which appears to imply that, after all, a physical theory of spacetime could have been discovered significantly earlier than 1908 (before the discovery of Maxwell's electrodynamics).

If Galileo had had students, the best of them might have indeed tried to employ his research method to his principle of relativity. Galileo demonstrated through thought experiments and almost certainly observed real experiments which proved that by performing experiments on a ship it is impossible to determine whether it is at rest or in motion. Galileo's imagined students would try to understand the physical meaning of his relativity principle, particularly why it is impossible to distinguish rest and uniform motion (motion with constant velocity). At that time, both rest and uniform motion had been regarded as different *absolute* states. The students would try to explore the internal logic of the principle of relativity to see whether that could shed some light on the apparently paradoxical principle, which states that what seems to be obviously different – absolute rest and absolute uniform motion – cannot be distinguished by experiment. Their analysis could easily lead them to two possible interpretations of the principle of relativity (absolute motion cannot be detected by mechanical experiments):

- There is absolute uniform motion but we cannot detect it.

- There is no absolute uniform motion which explains why we cannot detect it.

Then the students would explore the internal logic of each of these interpretations. As the analyses are given in [33, Chap. 3], here I will outline only the analysis of the second interpretation since it is this interpretation of Galileo's principle of relativity which logically contains the idea of spacetime.

At first sight the statement that that there is no absolute uniform motion makes no sense (for brevity, until the end of this chapter I will use the expression "absolute motion"). But its analysis quickly reveals that absolute motion is motion with respect to absolute space. Then the statement that there is no absolute motion implies that there is no absolute space. Further analysis reveals that the meaning of that is "there is no *one* space," because absolute space means *one* space which is shared by everyone. When Galileo's students overcome the initial shock from this logical results, they would realize that the second interpretation of Galileo's principle of relativity implies that there exist many spaces, not just one. Then in order to understand the meaning of such a conclusion they would scrutinize the concept of space and would realize that space is defined in terms of *simultaneity* – all space points existing at the moment 'now' (a space can be regarded as a global 'now'). Then, many spaces mean many sets of simultaneously existing space points. This in turn implies many 'nows' (since many spaces correspond to many different 'nows'). And finally, an analogy with many planes in a three-dimensional space may lead them to the insight that many spaces imply a four-dimensional space with time as the fourth dimension; only in such a space (spacetime) there may exist many spaces which correspond to different 'nows'.

Now I hope you understand in what sense the idea of spacetime is logically contained in Galileo's principle of relativity. If this imagined scenario had really happened either in the 18th or in the 19th century, the colossal illusion that reality is a three-dimensional world which exists at the present moment of time would have been identified long before it actually happened.

Galileo's groundbreaking discoveries gave birth to modern science, but our civilization could have also benefited enormously from the finest creation of his brilliant mind – his powerfully productive research strategy. If Galileo's research method of exploring the internal logic of ideas (particularly of fundamental ideas) had been developed into a special course that the university students from all over the world (at least science students) had to take before graduating, science would undoubtedly be more advanced now.[3]

---

[3]The most distinct feature of the *Institute for Foundational Studies "Hermann Minkowski"* (www.minkowskiinstitute.org) is the employment of a research strategy based particularly on Galileo's research method and on other successful methods behind the greatest discoveries in physics. In this sense the *Minkowski Institute* is without a counterpart in the world.

4 HOW RACING A LIGHT BEAM HELPED EINSTEIN TO IDENTIFY THE ILLUSION OF ABSOLUTE TIME

We saw in the previous chapter how Galileo's powerful research strategy led to three discoveries which profoundly deepened our understanding of reality. Of all scientists up to date it was only Einstein who adopted fully Galileo's research method and it is not surprising that he was also extremely successful. Like Galileo, Einstein analyzed many thought (Gedanken) experiments which played crucial role in the discovery of his theories of special and general relativity.

Albert Einstein
14 March 1879 – 18 April 1955

A thought experiment which Einstein first considered when he was sixteen and which tormented him for nine years, right to the discovery of special relativity in 1905, considered a situation in which one raced a light beam. Einstein imagined that he was moving as fast as a light beam and wanted to know how he would see the beam.

This thought experiment became a paradox for Einstein when he studied Maxwell's equations at the Polytechnic Institute in Zurich. According to Maxwell's electrodynamics light is an electromagnetic wave, i.e. an oscillating electromagnetic field. If one used the rule of velocity addition in classical mechanics, it follows that if Einstein and a light beam moved at the same velocity c, he would see a "frozen" electromagnetic wave, because he would be at rest with respect to it (exactly like two cars will be at rest relative to each other when they move at the same velocity on the highway).

What Einstein found paradoxical in this thought experiment is that according to the classical rule of velocity addition the light beam would appear as a "frozen" electromagnetic wave to an observer traveling at the same velocity c as the light beam, whereas according to Maxwell's electrodynamics the velocity of the light beam relative to the traveling observer should, in Einstein's view, be also c. In Maxwell's theory the velocity of light is a universal constant ($c = (\mu_0 \epsilon_0)^{-1/2}$, where μ_0 and ϵ_0 are electromagnetic constants – the permeability and permittivity of vacuum or empty space) and Einstein interpreted this to mean that it should be such a constant to all observers independent of their state of motion. Einstein trusted Maxwell's equations and that they "should hold also in the moving frame of reference" [34, p. 139], which meant for Einstein that if he travelled (almost) at the speed of light (relative, say, to Earth), a beam of light would still move away from him at velocity c, which is in Einstein's own words "in conflict with the rule of addition of velocities we knew of well in mechanics" [34, p. 139].

Einstein had an additional reason to believe that Maxwell's equations (describing a light beam as an electromagnetic wave propagating at speed c) should be equally valid for all observers in relative motion with constant velocity (i.e. for all inertial reference frames). If Maxwell's equations were interpreted to predict that the speed of light is c only with respect to space, then that would prove Newton's views of absolute space and absolute time. It would also present a huge challenge to Galileo's principle of relativity stating that absolute uniform motion (with constant velocity) cannot be detected with mechanical experiments. To see the challenge better, let us return to Einstein's thought experiment of racing a light beam. If the speed of the light beam were c only relative to the absolute space, then anyone traveling at c in the same direction as the beam would see it as a "frozen" electromagnetic wave and would prove that absolute uniform motion is detectable (when the observer is at rest relative to the absolute space light would travel at c relative to him, but if he sees a "frozen" light wave or the speed of the light beam with respect to him is not c he would have proof that he is uniformly moving relative to space and would be even able to calculate his absolute speed).

Einstein did not believe in absolute space, absolute motion, and any absolutes. Along with his studies in physics, he was reading books on philosophy (probably while skipping some lectures of his mathematics professor Hermann Minkowski) and those books made him a relativist. However, philosophical arguments do not work against physical arguments. Especially given the fact that the physical arguments that space was absolute and was some kind of a medium appeared irrefutable throughout the 19th century. In 1803 Thomas Young performed his famous double-slit experiment and proved that light was a wave. This experiment appeared to provide indirect proof that space is absolute for the following reason. As the double-slit experiment proved that light is a wave phenomenon and as, by its very nature, *a wave is a disturbance of a medium* it followed that space, where light propagates (e.g. when emitted from the Sun), is some kind of a medium (whose disturbances are observed as light). That medium was called ether or luminiferous (light-carrying) ether.

So, not only did Young's experiment prove that light was a wave, but it also implied that the ether was needed for the very existence of light.

The situation in fundamental physics in the 19th century had been really interesting and challenging. On the one hand, the *experimental* evidence captured in Galileo's principle of relativity clearly indicated that uniform motion could not be discovered with mechanical experiments. On the other hand, Young's double-slit *experiment* appeared to have proved that the luminiferous ether (i.e. absolute space) existed. The fact that experiments do not contradict one another clearly indicated that there was something wrong in this situation. An additional piece in the puzzle was the Michelson-Morley experiment, performed in 1887, which intended to detect the Earth's motion with respect to the ether by using *light* signals. It failed. The result was negative and it appeared as if the Earth was at rest with respect to the ether. Galileo's principle of relativity received an enormous boost – not only mechanical, but also electromagnetic (light) experiments cannot detect uniform motion.

The Michelson-Morley experiment could have provided physical arguments in support of Einstein's relativism, but, ironically, Einstein seemed to have been unaware of it [34, p. 172]

> In my own development, Michelson's result has not had a considerable influence. I even do not remember if I knew of it at all when I wrote my first paper on the subject (1905). The explanation is that I was, for general reasons, firmly convinced that there does not exist absolute motion and my problem was only how this could be reconciled with our knowledge of electrodynamics. One can therefore understand why in my personal struggle Michelson's experiment played no role, or at least no decisive role.

Einstein's "general reasons ... that there does not exist absolute motion" had been philosophical, but what ultimately helped him discover his special relativity before Lorentz, Poincaré, and Minkowski was a crucial physical argument – Planck's discovery that light behaved as a beam of discrete packets of electromagnetic energy. Einstein took very seriously Planck's hypothesis that light consisted of such quanta of electromagnetic energy and started to explore its internal logic. He analyzed conceptually and quantitatively (through calculations) a number of thought experiments involving interaction of electromagnetic radiation with mirrors and arrived at the conclusion that Planck's quanta must be real physical entities [35, p. 51]:

> If radiation were not subject to local fluctuations, the mirror would gradually come to rest, because, due to its motion, it reflects more radiation on its front than on its reverse side. However, the mirror must experience certain random fluctuations of the pressure exerted upon it due to the fact that the wave-packets, constituting the radiation, interfere with one another. These can be computed from Maxwell's theory. This calculation, then, shows that these pressure variations (especially in the case of small radiation-densities) are by no means sufficient to impart to the mirror the average kinetic energy $\frac{1}{2}(R/N)T$. In order to get this result one has to assume rather that there exists a second type of pressure variations, which can not be derived from Maxwell's theory, which corresponds to the assumption that radiation energy consists of indivisible point-like localized quanta of the energy $h\nu$ (and of momentum $(h\nu/c)$, (c = velocity of light)), which are reflected undivided. This way of looking at the problem showed in a drastic and direct way that a type of immediate reality has to be ascribed to Planck's quanta, that radiation must, therefore, possess a kind of molecular structure in energy, which of course contradicts Maxwell's theory.

If "a type of immediate reality has to be ascribed to Planck's quanta," then light possesses a *particle* aspect, which appears to indicate that *light is not necessarily a disturbance of the ether and does not need it for its existence and motion.* It was only Einstein who realized the significance of Planck's discovery for the struggle to reconcile classical mechanics with Maxwell's electrodynamics (whose contradiction is captured in Einstein's paradox of racing a light beam). This explains why it was Einstein who discovered special relativity.

Exploring the internal logic of Planck's idea helped Einstein not only in his efforts to resolve the contradiction between classical mechanics and Maxwell's electrodynamics, but also made him realize, in view of the contradiction between Planck's hypothesis and Maxwell's theory, that a universal principle which would "lead us to assured results" should be discovered [35, pp. 51-53]:

> Reflections of this type made it clear to me as long ago as shortly after 1900, i.e., shortly after Planck's trailblazing work, that neither mechanics nor electrodynamics could (except in limiting cases) claim exact validity. By and by I despaired of the possibility of discovering the true laws by means of constructive efforts based on known facts. The longer and the more despairingly I tried, the more I came to the conviction that only the discovery of a universal formal principle could lead us to assured results. The example I saw before me was thermodynamics. The general principle was there given in the theorem: the laws of nature are such that it is impossible to construct a *perpetuum mobile* (of the first and second kind). How, then, could such a universal principle be found? After ten years of reflection such a principle resulted from a paradox upon which I had already hit at the age of sixteen: If I pursue a beam of light with the velocity c (velocity of light in a vacuum), I should observe such a beam of light as a spatially oscillatory electromagnetic field at rest. However, there seems to be no such thing, whether on the basis of experience or according to Maxwell's equations. From the very beginning it appeared to me intuitively clear that, judged from the standpoint of such an observer, everything would have to happen according to the same laws as for an observer who, relative to the earth, was at rest. For how, otherwise, should the first observer know, i.e., be able to determine, that he is in a state of fast uniform motion?

> One sees that in this paradox the germ of the special relativity theory is already contained.

The universal principle which Einstein needed was the principle of relativity – Galileo's principle of relativity generalized to mean that uniform motion cannot be discovered, or expressed equivalently – physical phenomena look in the same way in all inertial reference frames (otherwise uniform motion could be discovered by the discrepancies in observing the physical phenomena in different frames).

The germ of special relativity may be contained in Einstein's paradox of racing a light beam, but he needed years of intense analysis and discussions with his friend Michele Besso before he was able to resolve the paradox. Here is how M. Kaku reconstructs Einstein's resolution of the racing light beam paradox [36]:

> One day around May of 1905, Einstein went to visit his good friend Michele Besso, who also worked at the patent office, and laid out the dimensions of the problem that had puzzled him for a decade. Using Besso as his favorite sounding board for ideas, Einstein presented the issue: Newtonian mechanics and Maxwell's equations, the two pillars of physics, were incompatible. One or the other was wrong. Whichever theory proved to be correct, the final resolution would require a vast reorganization of all of physics. Einstein went over and over the paradox of racing a light beam. ... They talked for hours, discussing every aspect of the problem, including Newton's concept of absolute space and time, which seemed to violate Maxwell's constancy of the speed of light. Eventually, totally exhausted, Einstein announced that he was defeated and would give up the entire quest. It was no use; he had failed.
>
> Although Einstein was depressed, his thoughts were still churning in his mind when he returned home that night. In particular, he remembered riding in a streetcar in Bern and looking back at the famous clock tower that dominated the city. He then imagined what would happen if his streetcar raced away from the clock tower at the speed of light. He quickly realized that the clock would appear stopped, since light could not catch up to the streetcar, but his own clock in the streetcar would beat normally.
>
> Then it suddenly hit him, the key to the entire problem. Einstein recalled, "A storm broke loose in my mind." The answer was simple and elegant: *time can beat at different rates throughout the universe, depending on how fast you moved.* Imagine clocks scattered at different points in space, each one announcing a different time, each one ticking at a different rate. ... This meant that events that were simultaneous in one frame were not necessarily simultaneous in another frame, as Newton thought.

If this reconstruction is correct and Einstein indeed identified the unsuspected-for-centuries illusion that time was absolute in this way, he got the right result for the wrong reason. His thought experiment "that the clock would appear stopped", "if his streetcar raced away from the clock tower at the speed of light" described a well-known *classical* effect involved in the Doppler effect – as the streetcar moves away from the clock tower each light signal from the clock will have to travel a greater distance (compared to the distance traveled by the previous signal) and for this reason will arrive a bit later. That this is a classical effect can be immediately seen if you assume that the streetcar approaches (not recedes from) the clock tower – the clock will appear to run faster (in relativity time only "slows down"). Unfortunately, Einstein was not very helpful in clarifying how he actually got the insight that time was not absolute, because in his recollection of the resolution of the paradox he talked again about a "connection between time and the signal velocity" [34, p. 139]:

> My solution was really for the very concept of time, that is, that time is not absolutely defined but there is an inseparable connection between time and the signal velocity. With this connection, the foregoing extraordinary difficulty could be thoroughly solved. Five weeks after my recognition of this, the present theory of special relativity was completed.

Once Einstein realized that observers in relative motion have different times, the paradox was finally resolved since each observer uses his own time in all measurements. For example, the velocity addition rule changes when it is taken into account that time is not absolute and every observer uses his time to calculate how velocities are added. As a result, as implied by Maxwell's electrodynamics, the velocity of light does turn out to be a universal constant – the same for all observers in relative motion – which means that a light beam will always move at velocity c relative to an observer no mater how fast the observer moves (that is, no one will ever be able to see a "frozen" electromagnetic wave).

In 1905 when Einstein published his paper on special relativity it appeared that the relativism of Berkeley and Mach had found their way in fundamental physics – even the name of Einstein's theory reflected those views. And indeed, Einstein declared that there is no absolute space (i.e. no ether) and therefore no absolute motion [37, p. 141]:

> The introduction of a "light ether" will prove superfluous, inasmuch as in accordance with the concept to be developed here, no "space at absolute rest" endowed with special

properties will be introduced, nor will a velocity vector be assigned to a point of empty space at which electromagnetic processes are taking place.

Let us summarize the two major reasons, which helped Einstein to reject the idea of absolute space and absolute motion and to arrive at special relativity before Lorentz and Poincaré. First, Einstein had been "for general reasons, firmly convinced that there does not exist absolute motion" [34, p. 172] mostly due to Mach's relativism. Second, Einstein had been impressed by the fact that the velocity of light is a fundamental constant in Maxwell's electromagnetic theory since it is expressed through the fundamental (universal) constants μ_0 and ϵ_0 (the permeability and permittivity of vacuum or empty space). Lorentz, Poincaré and the other physicists at the turn of the nineteenth and the twentieth centuries seemed to have seen in the definition of c in terms of μ_0 and ϵ_0 only further support for the luminiferous ether since μ_0 and ϵ_0 were interpreted as describing the properties of this light carrying medium. For Einstein, however, the fact that c is a fundamental (*universal*) constant was extraordinary since *a fundamental constant should be a constant in all inertial reference frames*. But if this is so, then c denotes the speed of light not in the ether (i.e. in the absolute space) but in *all* inertial reference frames. Whatever that meant for Einstein, it showed him that there was a serious problem with the concept of absolute space (if the absolute space existed, c should denote the speed of light with respect to it).

To make the above arguments against absolutism in physics even stronger, Einstein began his 1905 paper "On the Electrodynamics of Moving Bodies," which contained his special theory of relativity, with an example intended to demonstrate that the observed phenomena do not seem to support the concept of absolute space and therefore of absolute motion and absolute rest [37, p. 140]:

> It is known that Maxwell's electrodynamics – as usually understood at the present time – when applied to moving bodies, leads to asymmetries which do not seem to attach to the phenomena. Let us recall, for example, the electrodynamic interaction between a magnet and a conductor. The observable phenomenon depends here only on the relative motion of conductor and the magnet ... Examples of a similar kind, and the failure of attempts to detect a motion of the earth relative to the "light medium," lead to the conjecture that not only in mechanics, but in electrodynamics as well, the phenomena do not have any properties

corresponding to the concept of absolute rest, but that in
all coordinate systems in which the mechanical equations
are valid, also the same electrodynamic and optical laws
are valid ...

The advent of special relativity led not only to the relativization
of motion, but it also relativized simultaneity, time, length, and mass.
At first sight it seemed that the triumph of relativism in physics was
complete. However, the initial formulation of special relativity had the
major hallmark of relativism – assuming that there is nothing more
to phenomena (what we can observe) and ignoring the lessons from
the history of science (especially the way Galileo discovered inertial
motion and the principle of relativity) which demonstrate that the
phenomena are representing only "the surface of the world" since fundamental physical laws and profound features of the world manifest
themselves through the phenomena. So relativism was dealing only
with the manifestations of fundamental features of Nature effectively
postulating that there is nothing behind those manifestations and for
this reason not even trying to ask any deep questions about the nature
of reality.

Relativism is a sterile doctrine that is unable to decode the mes-
sages hidden in the phenomena and to lead to genuine and deep understanding of the world. Einstein was a happy exception. In his
1905 paper he used that doctrine only to reconcile Galileo's principle
of relativity and Maxwell's electrodynamics and after that he gradually abandoned it. However, Einstein used his enormous intellectual
potential to succeed in arriving at the theory of special relativity by
using such an unproductive doctrine. It took him years and a huge intellectual effort to resolve the apparent paradox of racing a light beam.

The major hallmark of relativism was clearly evident in the initial
formulation of special relativity – a number of apparently paradoxical results were simply postulated (on the basis of the experimental
evidence) without any attempt to provide deeper explanation. For
example, it was postulated that

- absolute space does not exist, whereas no one could deny that
  all objects exist and move in something that is called "the space"
  (absolute space means *one* space that is common to all observers)

- absolute motion does not exist, whereas objects move in "the
  space"

- absolute time does not exist; everyone was supposed to accept the fact that that mysterious result followed from the experiment and no more questions should be asked

- the speed of light was constant no matter how the source and the observer moved (no matter how fast one can chase a light beam it will always move away at speed c), and again everyone had to just accept this paradoxical result since it was what the experimental evidence showed

- the length of bodies along the line of their motion contracts, but the reality of that relativistic contraction was another mystery since no deformation and no force were causing the shortening of the length

- the time of a moving observer "slows down" but special relativity offers no explanation what that means and what causes it

- the traveling twin in the twin paradox effect is younger than his brother when they meet, but special relativity again offers no explanation.

As we will see in the next chapter (and as it was only mentioned in Chapter 1) Einstein's mathematics professor, Hermann Minkowski, gave a four-dimensional formulation of special relativity demonstrating that *relative* quantities can only exist as manifestations of an underlying *absolute* reality. For example, space and time (and therefore simultaneity as well) are relative because they are manifestations of an absolute four-dimensional world – spacetime. (Here is again the best example of why this is so – as Minkowski first noticed, observers in relative motion can have relative, that is, different, spaces *only* if the world is four-dimensional, i.e. only if spacetime is real; then these spaces are merely three-dimensional cross sections of spacetime. If spacetime were only an abstract construction and what existed were the ordinary three-dimensional space, obviously observers in relative motion could not have *different* spaces since just *one* space would exist.) All of the above questions have natural explanations in spacetime.

5   SPACETIME – MINKOWSKI STRANGE WORLD

I see it, but I don't believe it.
Georg Cantor [19]

Einstein had been enormously lucky to have such a great mind as Hermann Minkowski as his mathematics professor. It is now impossible to say whether Einstein had made a mistake when he had been skipping some of his lectures in order to read philosophical works on relativism. On the one hand, as Einstein himself explained and as the very names of his theories of special and general *relativity* indicate, those works helped him to discover the two theories of relativity. On the other hand, skipped mathematics lectures might have caused Einstein the difficulties he experienced to express his physical ideas about the nature of gravitation (that no gravitational force is responsible for the gravitational phenomena) into a mathematical language.

Fortunately, Einstein had as a classmate and friend Marcel Grossmann, who diligently attended all lectures and who later specialized in geometry to become a professor of descriptive geometry. He realized that Einstein's radically novel view of the nature of gravitation makes sense only in a non-Euclidean geometry where two particles only seem to interact through gravitational forces, whereas they actually move by inertia; the particles approach each other since there are no straight and parallel lines in a curved (i.e. non-Euclidean) space. Grossmann helped Einstein master the mathematical formalism of non-Euclidean geometries and

Hermann Minkowski
22 June 1864 – 12 January 1909

in 1913 both published a paper "Outline of a Generalized Theory of Relativity and of a Theory of Gravitation" [38] which paved the way toward Einstein's theory of general relativity.

However, as we will see in the next chapter, the revolutionary view that gravity is not a force as revealed by general relativity would have been impossible (as Einstein himself admitted) without Minkowski's four-dimensional (spacetime) formulation of special relativity. But before explaining how Minkowski did that, I think it is appropriate to outline the drama behind the discovery of special relativity and spacetime, which involved Einstein, Minkowski, and Poincaré.

The scientific life of Minkowski started perhaps in the best possible way. In April 1883 the French Academy granted the Grand Prize in Mathematics jointly to the eighteen year old Minkowski for his innovative geometric approach to the theory of quadratic forms and to Henry Smith. Thirteen years later, in 1896, Minkowski published his major work in mathematics *The Geometry of Numbers* [40].

By 1905 Minkowski was already internationally recognized as an exceptional mathematical talent. At that time he became interested in an unresolved issue at the very core of fundamental physics – at the turn of the nineteenth and twentieth century Maxwell's electrodynamics had been interpreted to show that light is an electromagnetic wave, which propagates in a light carrying medium (the luminiferous ether), but its existence was put into question since the Michelson and Morley interference experiments failed to detect the Earth's motion in that medium. Minkowski's involvement with the electrodynamics of moving bodies began in the summer of 1905 when he and his friend David Hilbert co-directed a seminar in Göttingen on the electron theory (dealing with the electrodynamics of moving bodies). Einstein's paper on special relativity was not published at that time; *Annalen der Physik* received the paper on June 30, 1905. Poincaré's longer paper "Sur la dynamique de l'électron" was not published either; it appeared in 1906. Also, "Lorentz's 1904 paper (with a form of the transformations now bearing his name) was not on the syllabus" [41].

Minkowski's student Max Born, who attended the seminar in 1905, recalled in 1959 what Minkowski had said during the seminar (quoted in [42]):

> I remember that Minkowski occasionally alluded to the fact that he was engaged with the Lorentz transformations, and that he was on the track of new interrelationships.

Again Born wrote in his autobiography about what he had heard from Minkowski after Minkowski's lecture "Space and Time" given on

September 21, 1908 [43]:

> He told me later that it came to him as a great shock when
> Einstein published his paper in which the equivalence of the
> different local times of observers moving relative to each
> other were pronounced; for he had reached the same con-
> clusions independently but did not publish them because
> he wished first to work out the mathematical structure in
> all its splendour. He never made a priority claim and al-
> ways gave Einstein his full share in the great discovery.

Einstein published his results before Minkowski and deservedly received the credit for the discovery of special relativity. The drama
of the discovery of spacetime, involving Minkowski and Poincaré, is
somewhat reversed. In his paper *Sur la dynamique de l'électron* [44]
published in 1906 (but received by *Rendiconti del Circolo matematico
Rendiconti del Circolo di Palermo* on July 23, 1905) Poincaré first
published the important result that the Lorentz transformations had
a geometric interpretation as rotations in what he seemed to have regarded as an *abstract* four-dimensional space with time as the fourth
dimension. But the credit for the discovery of the physical concept
of spacetime was given to Minkowski.[1] This needs some explanation
for two reasons. First, to explain why no scientific injustice was done.
Second, because the drama of the discovery of spacetime contains a
very important lesson for present and future scientists.

As we will see below, not only did Minkowski independently realize that the times of observers in relative motion should be treated
equally, but that discovery meant that different times imply different spaces as well, which is possible only in a four-dimensional space
(spacetime). So Minkowski realized the equivalence of the times of observers in relative motion independently of Einstein, and the concept
of spacetime independently of Poincaré. In fact, Minkowski did much
more than just arriving at the idea of spacetime. He fully developed
the four-dimensional physics based on the idea of spacetime which
he discovered on its own, but which was first published by Poincaré,
and he showed that spacetime is not just a convenient mathematical
space, but represents a real four-dimensional world. I think all these
facts justify giving the credit to Minkowski for his four-dimensional
physics, without denying that it was Poincaré who first published the
result that the Lorentz transformation can be regarded as a rotation
in a space of four dimensions (one of which is time).

---

[1] Ironically, not only the general public, but even students of physics appear to
believe that the concept of spacetime was introduced by Einstein.

In fact, there is a second reason for crediting Minkowski for the discovery of spacetime as a physical concept. Unlike Minkowski, Poincaré seems to have seen nothing revolutionary in the idea of a mathematical four-dimensional space as Damour remarked [45, p. 51]:

> although the first discovery of the mathematical structure of the space-time of special relativity is due to Poincaré's great article of July 1905, Poincaré (in contrast to Minkowski) had never believed that this structure could really be important for physics. This appears clearly in the final passage that Poincaré wrote on the question some months before his death [46].

Here is the final passage [46]:

> Everything happens as if time were a fourth dimension of space, and as if four-dimensional space resulting from the combination of ordinary space and of time could rotate not only around an axis of ordinary space in such a way that time were not altered, but around any axis whatever...
>
> What shall be our position in view of these new conceptions? Shall we be obliged to modify our conclusions? Certainly not; we had adopted a convention because it seemed convenient and we had said that nothing could constrain us to abandon it. Today some physicists want to adopt a new convention. It is not that they are constrained to do so; they consider this new convention more convenient; that is all. And those who are not of this opinion can legitimately retain the old one in order not to disturb their old habits. I believe, just between us, that this is what they shall do for a long time to come.

Poincaré even appeared to have thought that the spacetime convention would not be advantageous [47]:

> It quite seems, indeed, that it would be possible to translate our physics into the language of geometry of four dimensions. Attempting such a translation would be giving oneself a great deal of trouble for little profit, and I will content myself with mentioning Hertz's mechanics, in which something of the kind may be seen. Yet, it seems that the translation would always be less simple than the text, and that it would never lose the appearance of a translation, for

the language of three dimensions seems the best suited to the description of our world, even though that description may be made, in case of necessity, in another idiom.

Poincaré believed that our physical theories are only *convenient descriptions* of the world and therefore it is really a matter of *convenience* and *our choice* which theory we would use. As Damour stressed it [45, p. 52], it was

> the sterility of Poincaré's scientific philosophy: complete and utter "conventionality" ... which stopped him from taking seriously, and developing as a physicist, the space-time structure which he was the first to discover.

What makes Poincaré's failure to comprehend the profound physical meaning of the relativity principle and the geometric interpretation of the Lorentz transformations especially sad is that it is perhaps the most cruel example in the history of physics of how an inadequate philosophical position can prevent a scientist, even as great as Poincaré, from making a discovery. However, this sad example can serve a noble purpose. Science students and young scientists can study it and learn from it because scientists often think that they do not need any philosophical position for their research [48]:

> Scientists sometimes deceive themselves into thinking that philosophical ideas are only, at best, decorations or parasitic commentaries on the hard, objective triumphs of science, and that they themselves are immune to the confusions that philosophers devote their lives to dissolving. But there is no such thing as philosophy-free science; there is only science whose philosophical baggage is taken on board without examination.

Let us now see how Minkowski arrived independently at the two important results – the equivalence of the times of observers in relative motion and the fact that that equivalence implies a four-dimensional space with time as the fourth dimension. In this way we will see that Born's recollections only confirm what follows from a careful study of the results of Minkowski's publications.

Shortly after Einstein published his special relativity in 1905, Minkowski, delivered three lectures – November 5, 1907, December 21, 1907, and September 21, 1908 – (before his untimely at the age of 44 departure from this strange world on January 12, 1909) which dramatically changed the way we understand the world [12]. He realized that the experimental evidence supporting Galileo's principle of relativity (absolute motion with constant velocity cannot be discovered through mechanical experiments), including the failed attempts by Michelson and Morley to detect the Earth's motion by electromagnetic experiments involving light beams (since light is an electromagnetic wave), carried a stunning message – all mechanical and electromagnetic experiments to discover uniform motion with respect to the absolute space (i.e. absolute motion) fail since observers in relative motion have different spaces and times, which is possible in a four-dimensional world whose fourth dimension is formed by *all* moments of time.

Most probably, Minkowski's mathematical way of thinking helped him to take seriously an abstract mathematical time t' introduced by Lorentz in his *mathematical* description of the negative result of the Michelson and Morley experiment. Lorentz formally introduced t', calling it the *local time* of a moving observer, in addition to the real time t of a stationary observer. As a physicist Lorentz was putting physics first, believing that it was a physical fact that there existed just one time (thus falling victim to the illusion of absolute time), whereas for Minkowski *both* times t and t' successfully described a real physical situation and therefore should be treated equally. I think this appears to have been the line of thought which led Minkowski to the realization of the equality of the times t and t' independently of Einstein. The fact that in his second lecture (on December 21, 1907) Minkowski presented the full-blown mathematical formalism of the spacetime formulation of special relativity (which might have taken years to develop) is another piece of evidence that gives weight to Minkowski's words that he had reached the conclusion of the equality of t and t' independently but did not publish it "because he wished first to work out the mathematical structure in all its splendour."

As a mathematician it may have been easier for Minkowski (than for Einstein) to postulate that the times t and t' are equivalent and to explore the consequences of such a hypothesis. The mathematical way of thinking surely had helped Minkowski to realize that if two observers in relative motion have different times they necessarily must have different spaces as well (since space is perpendicular to time), which is impossible in a three-dimensional world, but is almost self-evident in a four-dimensional world with all moments of time as the fourth dimension. Here is how Minkowski in his own words at his lecture "Space and Time" explained how he had realized the profound *physical meaning of the relativity principle* – that the world is four-dimensional. In the case of two inertial reference frames in relative motion along their x-axes [12, p. 114] one can call t' time, but then must necessarily, in connection with this, define space by the manifold of three parameters x', y, z in which the laws of physics would then have exactly the same expressions by means of x', y, z, t' as by means of x, y, z, t. Hereafter we would then have in the world no more *the* space, but an infinite number of spaces analogously as there is an infinite number of planes in three-dimensional space. Three-dimensional geometry becomes a chapter in four-dimensional physics.

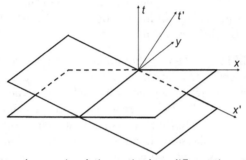

Figure 5.1: If two observers in relative motion have different times they unavoidably have different spaces as well. As the observers move along their x—axes, they share the same y and z axes. Here the third (z) dimension is suppressed.

Minkowski's papers, especially his "Space and Time," reveal the following logic behind his discovery that reality is an absolute four-dimensional world, which is strikingly different from the logic behind Einstein's and Poincaré's discoveries. This is another piece of evidence that Minkowski did discover independently spacetime and its four-dimensional physics, but, unfortunately, did not publish any preliminary results during the time he was developing the novel mathematical spacetime formalism.

- In order to explain the negative result of the Michelson and Morley experiment that light propagates on Earth as if the Earth were at rest relative to the ether (i.e. that the Earth's uniform motion through the ether cannot be detected with light signals), Lorentz introduced a mathematical time t' assigned to the moving Earth in addition to the real absolute time t associated with the absolute space (the ether).

- As from a *mathematical* point of view t and t' have *the same status in the mathematical descriptions* of the phenomena involved in the Michelson and Morley experiment, Minkowski assumed that they both represent physical quantities. This meant that observers in relative motion (e.g. one "stationary" and one on Earth) have different times. Then Minkowski realized that, taken seriously, Lorentz' hypothesis explained not only the null result of the Michelson and Morley experiment, but all the experimental evidence that uniform motion cannot be detected, or, what is the same, the fact that all physical laws are the same in all inertial reference frames (because if they were not the same, the discrepancies would prove that uniform motion can be discovered). In other words, Lorentz' hypothesis explained the principle of relativity – all physical phenomena are the same in all inertial reference frames since all inertial observers describe them in the *same* way by their own times.

- As seen in Fig. 5.1 the fact that observers in relative motion have different times (i.e. relative times) implies that they have different spaces (i.e. relative spaces) as well. Then the explanation of the experimental evidence reflected in the principle of relativity becomes complete (in three-dimensional language) – physical phenomena are the same in all inertial reference frames because every inertial observer describes the phenomena in *exactly the same way* in his own reference frame (i.e. in terms of his own space and time) in which he is *at rest*. For example, the answer to the question of the failure of the Michelson and Morley experiment to detect the motion of the Earth appears obvious – the Earth is at rest with respect to its space and therefore not only the Michelson and Morley but any other experiments would confirm this state of rest. As every observer always measures the velocity of light in his own (rest) space and by using his own time, the velocity of light is the same for all observers. Einstein postulated the principle of relativity and the constancy of the speed of light. Minkowski explained them.

- Minkowski remarked that "The concept of space was shaken neither by Einstein nor by Lorentz" [12, p. 117] perhaps trying to imply that it was their failure to notice that a changed concept of time leads to a changed concept of space as well, which prevented them from realizing the physical meaning of the relativity of time. Indeed, *taken alone*, it is unclear what the meaning of the postulate that time is relative is. But when it is taken explicitly into account that relative time implies relative space, i.e. that many times imply many spaces, a basic geometrical imagination immediately demonstrates (as seen in Fig. 5.1) that many spaces (and many times) imply a four-dimensional space with time as one of the dimensions "analogously as there is an infinite number of planes in three-dimensional space." Exactly as no relative (many) planes and no relative (many) $z-$axes (associated with the planes) would be possible if the three-dimensional space did not exist, no relative spaces and no relative times would be possible, if spacetime did not exist. It is this basic geometric fact which demonstrates that *no relative quantities are possible without an underlying absolute reality.* In other words, the very existence of relative quantities reveals the existence of an underlying absolute reality, because the relative quantities are manifestations of that absolute reality. That is why Minkowski found that the relativity principle, which Einstein used as the first postulate of his theory of special relativity (the second postulate being the constancy of the speed of light), did not adequately represent what the experimental evidence tells us about the world and noted that "I think the word *relativity postulate* used for the requirement of invariance under the group G_c is very feeble. Since the meaning of the postulate is that through the phenomena only the four-dimensional world in space and time is given, but the projection in space and in time can still be made with certain freedom, I want to give this affirmation rather the name *the postulate of the absolute world*" [12, p. 117].

Understanding the logic behind Minkowski's discovery of spacetime makes it possible to understand his excitement when on September 21, 1908 in his lecture "Space and Time" (whose most important transparency is shown in Fig. 5.2) he announced the revolutionary view of space and time deduced from experimental physics by successfully decoding the profound message hidden in the failed experiments to discover absolute motion [12, p.111]:

> The views of space and time which I want to present to you arose from the domain of experimental physics, and therein lies their strength. Their tendency is radical. From now onwards space by itself and time by itself will recede completely to become mere shadows and only a type of union of the two will still stand independently on its own.

Minkowski's excitement appears to have been caused not only by the discovery itself, but perhaps even more by the unprecedented explanatory power of the new four-dimensional view of the world. In

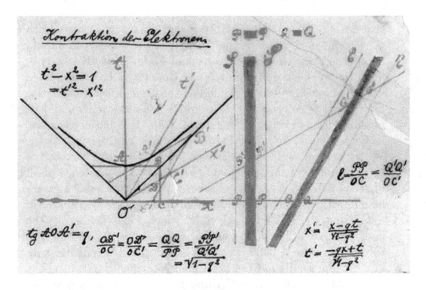

Figure 5.2: The transparency which Minkowski used at his lecture in Cologne on September 21, 1908. It shows Fig. 1 in his paper " Space and Time." Source: Cover of *The Mathematical Intelligencer*, Volume 31, Number 2 (2009).

addition to revealing the profound message carried by the failed experiments to discover absolute motion (i.e. explaining the physical meaning of the relativity principle since that experimental evidence was merely postulated in the relativity principle), the spacetime concept explained everything that was determined by experiment but was simply postulated due to the lack of any explanation before Minkowski's revolution. For example, as there is "in the world no more *the* space, but an infinite number of spaces" it becomes clear why there does not exist *one* (absolute) space and therefore why there does not exist absolute motion (motion relative to the absolute space). By the same reason, time is not absolute since observers in relative motion can choose their time axes along different directions in spacetime. As we will see below the physical meaning of the kinematical relativistic effects is fully revealed by the spacetime concept.

Minkowski called the four-dimensional world (spacetime) the absolute world (since it is the *same* for all observers) or, shortly, the world. He named all its points worldpoints; we now also call them events (an event is defined as a particle or a space point at a given moment of time). Since *spacetime contains all moments of time at once* the whole history in time of a particle is entirely given in spacetime. Minkowski called the "line of life" of such a particle a worldline. He generalized Newton's first law (of inertia) by pointing out that a

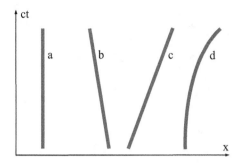

Figure 5.3: Worldlines of three particles (a, b, c) moving with constant relative velocities and of an accelerating particle (d). Three observers at rest with respect to particles a, b, and c can choose their time axes along the worldlines of these particles.

free particle, which moves by inertia, is a straight worldline in the absolute world (spacetime) and pointed out that an accelerating particle is represented by a curved worldline. Here is how he described the three states of motion of a particle (corresponding to the worldlines a, c, and d in Fig. 5.3) [12, p. 115]:

> a straight worldline parallel to the t-axis corresponds to a stationary substantial point, a straight line inclined to the t-axis corresponds to a uniformly moving substantial point, a somewhat curved worldline corresponds to a non-uniformly moving substantial point.

In ordinary space all directions are equivalent due to the fact that all dimensions of space are of the same kind – spatial dimensions. However, in spacetime there are three spatial dimensions and one temporal (time) dimension. As a result, not all directions are equivalent. As shown in Fig. 5.4 there are three directions (and therefore three kinds of length) in spacetime – *timelike* directions (e.g. OA) along which the worldlines of particles (moving at speed smaller than that of light c) are extended; *lightlike* directions (e.g. OB) along which the worldlines of light signals are extended; and *spacelike* directions (e.g. OC) along which the spaces of different observers can be chosen. All lightlike worldlines, representing all light signals emitted from event O, form the upper cone in Fig. 5.4. It is called the *future light cone* since it contains all lightlike worldlines of all light signals emitted from O and propagating in all directions "toward the future." The lower cone in Fig. 5.4 is called the *past light cone* since it comprises all lightlike worldlines of all light signals coming from all directions of space (from past moments) and converging toward O.

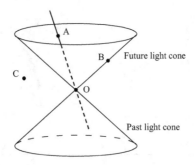

Figure 5.4: There are three directions in spacetime – timelike (OA), lightlike (OB), and spacelike (OC). The lightlike worldlines of all light signals, emitted at past moments of time and converging from all directions toward O, form the past light cone. The lightlike worldlines of all light signal, emitted from O and propagating in all direction, form the future light cone.

In spacetime where the whole history of the three-dimensional world is given *en bloc* (and for this reason spacetime is sometimes called a *block universe*) there are only worldlines and no interactions.[2] This fact led Minkowski to the conclusion that the whole physics is in fact spacetime geometry [12, p. 112]:

> The whole world presents itself as resolved into such world-lines, and I want to say in advance, that in my understanding the laws of physics can find their most complete expression as interrelations between these worldlines.

Here are several examples of how Minkowski's understanding of physics as spacetime geometry works. Since a straight worldline represents inertial motion it immediately becomes clear why experiments have always failed to distinguish between a state of rest and a state of uniform motion – in both cases a particle is a *straight* worldline as seen in Fig 5.3 (worldlines a and b) and there is clearly no distinction between two straight lines. In the figure the time axis of the reference frame is along worldline a and the particle represented by this world-line appears to be at rest in the reference frame. If the time axis of another reference frame is chosen along worldline b, the particle represented by that worldline will appear to be at rest in the new reference frame, whereas the first particle (represented by worldline a) will appear to be uniformly moving with respect to the second particle (since

---

[2]This is the reason why in his original lectures notes on general relativity [49] R. Geroch stressed that "There is no dynamics in spacetime: nothing ever happens there. Spacetime is an unchanging, once-and-for-all picture encompassing past, present, and future."

worldline a is inclined to the new time axis, i.e. inclined to worldline b). Minkowski seems to have been impressed by this elegant explanation of the experimental fact that rest and uniform motion cannot be distinguished (which is an explanation of the relativity principle) that he decided "to introduce this fundamental axiom: *With appropriate setting of space and time the substance existing at any worldpoint can always be regarded as being at rest* [12, p. 115]. Therefore, inertial motion turns out to be absolute – in all reference frames the worldline of an inertial particle (moving with constant velocity) is straight.

Minkowski also showed that another centuries-old puzzle – acceleration is experimentally detectable – also had an elegant explanation in terms of spacetime geometry: an accelerating particle is a *curved* worldline in spacetime (worldline (d) in Fig. 5.3). He expressed this observation by stressing that "Especially the concept of *acceleration* acquires a sharply prominent character" [12, p. 117]. Therefore, the fact that acceleration is absolute (since it can be discovered by experiment) finds a natural explanation – in all reference frames the worldline of an accelerating particle is curved (i.e. deformed).

The *absoluteness* (frame-independence) of acceleration and inertial motion is reflected in the *curvature* of the worldline of an accelerating particle and the *straightness* of the worldline of a particle moving by inertia, respectively, which are *absolute geometrical properties* of the particles' worldlines. Then it becomes clear that *the acceleration of a particle is absolute not because the particle accelerates with respect to some absolute space, but because its worldline is curved*, which is a frame-independent geometrical fact. This deep understanding of inertial and accelerated motion in terms of the *shape* of particles' worldlines and *with no reference to space* nicely explains the apparent paradox that seems to have tormented Newton the most – both an inertial and an accelerating particle (appear to) move *in* space, but only the accelerating particle resists its motion.

Had Minkowski lived longer he almost certainly would have noticed that his program of regarding physics as geometry of spacetime would have produced another amazing result – as an accelerating particle *resists* its acceleration and as the worldtube[3] of such a particle is *deformed* it appears natural to assume that the particle's curved worldtube *statically* resists its deformation exactly like an ordinary (three-dimensional) curved rod statically resists its deformation. Then the particle's inertia, i.e. its resistance to its acceleration, can be viewed as originating from a four-dimensional stress in the deformed worldtube

---

[3]When we talk about real particles, which are spatially extended, it is more appropriate to use the term worldtube instead of worldline.

of the particle [51, Chap. 5]. It turns out that the *static* restoring force existing in the deformed worldtube of an accelerating particle does have the form of the inertial force with which the particle resists its acceleration [33, Chap. 9]. Minkowski would have been truly thrilled to realize that inertia is another manifestation of the reality of the four-dimensional world.

This explanation of inertia nicely completes the understanding of the absoluteness of accelerated and inertial motion. Accelerated motion is absolute due to an absolute (frame-independent) geometrical feature of the worldtube of an accelerating particle – its curvature. Now that absolute geometrical feature was linked to an absolute physical feature – the deformation of the worldtube of an accelerating particle (since a curved worldtube in flat spacetime means a deformed worldtube), which is accompanied by the static resistance of the worldtube caused by its deformation. It is this resistance to acceleration that makes accelerated motion detectable. Inertial motion is not detectable since the worldtube of any particle moving by inertia is straight, i.e. not deformed, and for this reason such a particle offers no resistance to its motion with constant velocity (constant speed and constant direction).

Since inertial mass is defined as the measure of the resistance a body offers to its acceleration it is now clear that a body's inertial mass is linked to the four-dimensional stress in the body's deformed worldtube.[4] I think this situation provides additional insight into the nature of relativistic mass especially in view of some recent attempts to deny the relativistic increase of mass. Paying particular attention to the main feature of inertia (*resistance*) captured in the accepted since Newton definition of mass demonstrates that relativistic mass is indeed one of the important results of special relativity – *as inertial mass is the measure of the resistance a body offers to its acceleration and as its acceleration is different in different inertial reference frames, the body's inertial mass cannot be the same in all frames.* In the inertial reference frame in which a body is at rest its mass is called rest (or proper) mass. It is an invariant, but its mass is greater in all inertial reference frames in relative motion with respect to the body. This relativistic generalization of the concept of mass is analogous to the relativistic generalization of the concept of time – proper time (the time between two readings of a clock in its rest frame) is an invariant

---

[4] As an elementary particle is not a worldline in spacetime its inertia appears to originate from a similar mechanism involving deformation at the quantum scale – from the *distorted* fields which mediate the particle's interactions [33, Chap. 9]; the particle's fields are deformed by its acceleration.

but coordinate time (the time between the same readings of the clock measured in the inertial frames moving relative to the clock) is greater; for a more detailed discussion of relativistic mass see [33, pp. 114-116].

It is evident that the spacetime explanation of the origin of inertia makes sense only if the worldtube of an accelerating particle is a *real* four-dimensional object. This explanation of inertia shows that even for practical reasons the implications of the reality of spacetime should be thoroughly explored because, for example, if we understand the origin of inertia we will be in a position to determine whether inertia can be controlled. But, obviously, the issue of the reality of spacetime (i.e. that the world is four-dimensional) should be resolved first.

We already saw that the experimental evidence behind the relativity principle could not be explained if the world were three-dimensional, i.e., if the absolute four-dimensional world (spacetime) were just a mathematical abstraction. This has been realized not only by Minkowski, but also by the majority of physicists who specialized in spacetime physics as seen from what some of them wrote (quoted at the end of Chapter 1). Here I will only quote again Eddington's direct answer to the question of the reality of spacetime since it was given in 1921 not long after Minkowski's discovery (Eddington calls the spaces and times of observers in relative motion *fictitious* since such spaces do not represent anything real because they are *imaginary* three-dimensional cross-sections of spacetime, exactly like the xy planes of different coordinate systems are imaginary two-dimensional cross-sections of space; analogously, the observers' times are fictitious because they can be chosen along the worldline of *any* uniformly moving particle, exactly like the z directions in space can be freely chosen[5]) [50, p.803]:

> It was shown by Minkowski that all these fictitious spaces and times can be united in a single continuum of four dimensions. The question is often raised whether this four-dimensional space-time is real, or merely a mathematical construction; perhaps it is sufficient to reply that it can at any rate not be less real than the fictitious space and time which it supplants.

Let us now examine what seems to be the most spectacular proof of a world view in the history of science. We owe both the world view

---

[5]This comparison can be made rigorous if it is noted that the z directions in space can be chosen in any direction without any restriction, whereas the time axes in spacetime can be chosen only along timelike worldlines representing particles. In the same way the xy planes in space can be freely chosen without restrictions, whereas the spaces of observers in spacetime should be along spacelike directions.

(the spacetime world view) and the proof to Minkowski – the experiments which confirmed the kinematical relativistic effects would be
impossible if spacetime (i.e. the four-dimensional world) did not exist;
stated another way, the relativistic experiments would be impossible
in a three-dimensional world.

Let us start with relativity of simultaneity or Minkowski's version
of this consequence of special relativity – observers in relative motion
have different spaces. As discussed in Chapter 1, a space constitutes a
class of simultaneous events and therefore, having different spaces, observers in relative motion have different classes of simultaneous events
(relativity of simultaneity). If the world were three-dimensional, space
would be absolute since there would exist only one space that would
be shared by all observers in relative motion. As space is a single class
of simultaneous events, absolute space implies absolute simultaneity
and therefore absolute time as well. All this is in a clear contradiction
with special relativity.

Two things about relativity of simultaneity were mentioned in
Chapter 1. First, the proof in the above paragraph, that no relativity of simultaneity is possible in a three-dimensional world, is valid
only if existence is absolute. Although Minkowski's arguments clearly
demonstrated that no relative quantities would be possible if an underlying absolute reality did not exist, we will see below that the idea to
relativize existence (to preserve the three-dimensionality of the world)
contradicts the experiments that confirmed the twin paradox effect.
Second, relativity of simultaneity has never been directly tested experimentally, but length contraction and time dilation, which are specific
manifestations of relativity of simultaneity, have been experimentally
confirmed. Therefore, relativity of simultaneity, taken even alone, is
sufficient to prove the reality of spacetime.

A more detailed argument is Minkowski's explanation of the deep
physical meaning of length contraction of two bodies in relative motion. The essence of his explanation is that length contraction is a manifestation of the reality of the bodies' worldtubes (Minkowski called
them strips). This can be best understood from Fig. 1 of Minkowski's
paper "Space and Time" (the right-hand part of which is reproduced
in Fig. 5.5 here) – length contraction would be *impossible* if the worldtubes of the two bodies, represented by the vertical and the inclined
strips in Fig. 5.5, did not exist and were nothing more than abstract
geometric constructions. To see this even more clearly consider only
the body represented by the vertical worldtube. The three-dimensional
cross-section PP, resulting from the intersection of the body's worldtube and the space of an observer at rest with respect to the body, is

the body's proper length. The three-dimensional cross-section $P'P'$, resulting from the intersection of the body's worldtube and the space of an observer moving with respect to the body, is the relativistically contracted length of the body measured by that observer[6]. Minkowski stressed that "This is the meaning of the Lorentzian hypothesis of the contraction of electrons in motion" [12, p.116].

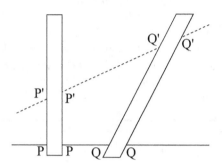

Figure 5.5: The right-hand part of Minkowski's Fig. 1

By demonstrating that the length contraction of a body is a manifestation of the reality the body's worldtube (and therefore of the reality of the absolute four-dimensional world) Minkowski also showed that this effect involves no deformation and no force causing the shortening of the body's length; as seen in Fig. 5.5 the contracted body measured by the observer moving relative to the body is simply *a different cross-section* $P'P'$ of the body's worldtube, which is shorter that the cross-section PP measured by the observer at rest with respect to the body. So, the length contraction effect is a nice illustration of the essence of Minkowski's four-dimensional physics – that *four-dimensional physics is spacetime geometry.*

It should be stressed that *the worldtube of the body must be real in order that length contraction be possible* because, while measuring the *same* body, the two observers in relative motion measure *two* three-dimensional bodies represented by the cross-sections PP and $P'P'$ in Fig. 5.5. This is not so surprising when one takes into account relativity of simultaneity and the fact that a spatially extended three-dimensional object is defined in terms of *simultaneity* – all parts of a body taken *simultaneously* at a given moment. If the worldtube of the body were an abstract geometric construction and what existed

---

[6]The cross-section $P'P'$ only appears longer than PP because a fact of the pseudo-Euclidean geometry of spacetime is represented on the Euclidean surface of the page.

were a single three-dimensional body (a single class of simultaneous events) represented by the proper cross-section PP, both observers would measure the *same* three-dimensional body of the *same* length, i.e. the *same* class of simultaneous events, which means that simultaneity would be absolute.

Length contraction was tested experimentally, along with time dilation, by the muon experiment in the muon reference frame (see for instance [69]).

Length contraction of a body, also taken even alone, is sufficient to prove that this relativistic effect would be impossible if the body's worldtube were not real, that is, if the world were three-dimensional. Due to this fact, let us examine a thought experiment which visualizes Minkowski's explanation of length contraction.

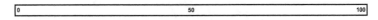

Figure 5.6: An ordinary meter stick.

The thought experiment clearly demonstrates that length contraction of a meter stick would be impossible if the meter stick existed as a three-dimensional body (not a worldtube). An ordinary meter stick (Fig. 5.6) is at rest with respect to an observer A. What is shown in Fig. 5.6 is what we perceive and take for granted that it is what really exists. According to Minkowski, however, the meter stick exists equally at all moments of its history and what is ultimately real is the worldtube of the meter stick as shown in Fig. 5.7.

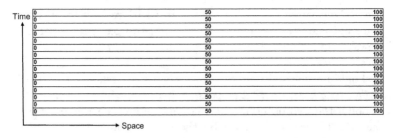

Figure 5.7: The worldtube of the meter stick.

Assume that another meter stick at rest in another observer's (observer B's) reference frame moves relative to the first one at a distance 1 mm above it. Let us assume that at the event M the middle point of B's meter stick is instantaneously above the middle point of A's meter stick. Lights are installed inside A's meter stick, which change their color *simultaneously* at every instant in A's frame. At the event of the meeting M all lights are white in A's frame. At all previous moments all lights were bright gray. At all moments after the meeting all lights were dark gray. When A and B meet at event M this event is present for both of them. At that moment all lights of A's meter stick will be *simultaneously* white for A. In other words, the present meter stick for A is white (that is, all parts of A's meter stick, which exist *simultaneously* for A, are white), All moments before M when all lights of the meter were bright grey are past for A, whereas all moments when the meter stick will be dark grey are in A's future.

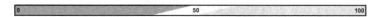

Figure 5.8: Relativistically contracted meter stick measured by observer B.

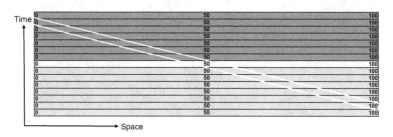

Figure 5.9: The worldtube of the meter stick with different colors.

Imagine that B's meter stick contains cameras, instead of lights, at every point along its length. At the event of the meeting M all cameras take snapshots of the parts of A's meter stick which the cameras face. All snapshots are taken *simultaneously* in B's reference frame. Even without looking at the pictures taken by the cameras it is clear that not all pictures will show a white part of A's meter stick, because *what is simultaneous for A is not simultaneous for B*. When the picture of A's meter stick is assembled from the pictures of all cameras it would show two things as depicted in Fig 5.8 – (i) A's meter stick photographed by B is shorter, and (ii) only the middle part of the picture of A's meter stick is white; half is bright grey and the other half is dark grey. So what is past (bright grey), present (white), and future (dark grey) for A exists *simultaneously* as present for B. But this is only possible if the meter stick is the worldtube as shown in Fig. 5.9. The instantaneous space of B corresponding to the event M intersects the worldtube of the meter stick at an angle and the resulting three-color "cross section" is what is measured by B – a different three-dimensional meter stick,

84

which is shorter[7] than the meter stick measured by A.

It should be emphasized again that no length contraction would be possible if the meter stick's worldtube did not exist as a four-dimensional object. Otherwise, if the meter stick were a three-dimensional object, both observers would measure the *same* three-dimensional meter stick (the same set of *simultaneously* existing parts of the meter stick), which would mean that the observers would share the same (absolute) class of simultaneous events in a clear contradiction with relativity.

Now let us see that time dilation would also be impossible if the worldtubes of two digital clocks A and B in relative motion, shown in Fig. 5.10, did not exist and the clocks were the familiar three-dimensional bodies.

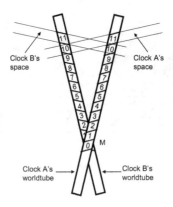

Figure 5.10: Reciprocal time dilation of two digital clocks

When the clocks meet at event M their readings are set to zero. Let two observers A and B be at rest with respect to clocks A and B, respectively. The two observers are performing identical experiments, which last ten seconds. The duration of the experiments is measured by the corresponding clock – the experiment carried out by observer A is measured by clock A, whereas B's experiment is measured by clock B. The time measured by the same clock is called *proper time* in relativity. In terms of spacetime proper time is a length of a timelike worldline. Indeed, as seen in Fig. 5.10 the worldtubes of clocks are "time rulers" in spacetime since their length is measured in seconds,

---

[7]In Fig. 5.9 the inclined " cross section," which represents the different three-dimensional meter stick measured by B, appears longer, not shorter, because a fact in the pseudo-Euclidean geometry of spacetime is represented on the Euclidean surface of the paper.

for example.

Each observer would like to determine what is the duration of the experiment carried out by the other observer. This can be done in the following way. Observer A determines which reading of the A clock is simultaneous with the 10th second of B's clock (the end of the experiment performed by B). As seen in Fig. 5.10 clock A's instantaneous space intersects the worldtube of clock B at its 10th second and the worldtube of clock A at the 11th second, which means that B's 10th second is simultaneous with A's 11th second. This relativistic result is called *time dilation* since as measured by A's clock the experiment performed by B takes *longer* time – 11 seconds. As seen in the figure the spacetime explanation of this effect is perfectly clear – it is just spacetime *geometry*. There is nothing mysterious in the fact that for observer A the experiment performed by B takes 11 seconds; simply A's instantaneous space intersects the worldtubes of the two clocks in such a way that it "cuts out" different segments of the clocks' worldtubes (no one finds anything mysterious when looking at the same figure, but drawn on ordinary Euclidean surface). The time dilation effect becomes mysterious and confusing only when one inexplicably starts talking about slowing down of B's time. There are only worldlines in spacetime; there is no flow of time there that can be slowed down.

If observer B decides to determine the duration of the experiment carried out by A, he follows the same procedure. As seen in Fig. 5.10 B's instantaneous space intersects clock B's worldtube at the 11th second and the worldtube of clock A – at the 10th second. B concludes that by his time the experiment at A takes 11 seconds.

Now the crucial part – what is the message implied by the time dilation effect? Minkowski made it clear – relative quantities imply some underlying absolute reality. The relativity of the two clock's times implies that their worldtubes are real four-dimensional objects. To see that, assume the opposite – that the clocks' worldtubes are nothing more than abstract geometrical constructions. Then the two clocks are the familiar digital clocks, which means that each of them exists only at their present moment. If this is so – if the clocks existed *only* at their 10th seconds – no time dilation would be possible as can be seen in Fig. 5.10 (if you imagine that of each worldtube only the 10th second exist).

The time dilation effect has been probably tested experimentally the most. It is sufficient to mention that it is used in the Global Positioning System. Now I hope you understand why the relativistic experimental evidence proved a whole worldview – that experimental evidence would be impossible if spacetime (the four-dimensional world) did not exist; stated another way, the relativistic experimental evidence is impossible in a three-dimensional world.

As discussed above and as this effect (along with length contraction) is a specific manifestation of relativity of simultaneity the above conclusion is true if existence is absolute. The only way to preserve some version of the view that reality is a three-dimensional world is to relativize existence. To see that even a relativized three-dimensionalism or presentism contradicts the relativistic experimental evidence, let us consider the twin paradox effect which is an *absolute* effect with no relativity of simultaneity involved. This absolute effect is depicted in Fig. 5.11 where the worldtubes of twins A and B are shown. Initially A and B are at rest with respect to each other – their worldtubes are parallel before the event D at which twin B departs, and after turning back at event T meets A again at the event M. Twin A's worldtube is a straight line, which means that he does not accelerate.

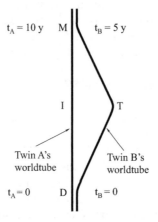

Figure 5.11: The twin paradox effect

In Euclidean geometry, the straight line is the shortest distance between two points. In the pseudo-Euclidean geometry of Minkowski spacetime, however, the straight worldline is the longest among all worldlines connecting two events. As the proper time of an observer is measured along his worldtube, each of the twins measures his elapsed proper time along his worldtube. The time that has elapsed between events D and M according to twin A is greater than the time as measured by twin B – A's worldtube between D and M is longer than B's worldtube between the same events. (In Fig. 5.11, it is the opposite since the diagram is drawn in the Euclidean geometry of the page.)

When A and B meet at M, it is an absolute fact that, for example, five years have passed for B and ten years for A. The time difference between the twins' clocks is an absolute effect because the twins directly compare their clocks at M. Therefore no relativity of simultaneity is responsible for this effect and no relativization of existence is necessary to explain that difference.

Let us now see that the twin paradox effect is possible only in a four-dimensional world, in which the twins' worldtubes are real four-dimensional objects. This can be seen by using the same approach as in the time dilation effect above – start from the opposite view that the twins' worldtubes are not real, and that the twins are three-dimensional objects existing only at their moment 'now'. In such a case both A and B should exist at the event M – otherwise what kind of a meeting would it be if they were not both present there? A and B can explain the time difference of five years *only* if they assume that B's time 'slowed down' during his journey. As the only difference in the states of motion of A and B is the acceleration that B has undergone during his journey, it follows that it is B's acceleration that should be responsible for the time difference. However, according to the so-called 'clock hypothesis' and the experiments which confirm it (see, for example, [53]) the rate of an ideal clock is unaffected by its acceleration [54, p. 164], [55], [56]. As nothing slows down twin B's time there would be no time difference in the twins' clocks at event M. Therefore, the twin paradox effect is impossible if the twins' worldtubes were not real and the twins existed only at their present moments.

Here is a second and independent argument which demonstrates the same. As proper times are not relativistically dilated (proper time is an invariant), the proper times of observers in relative motion (existing at their present moments as three-dimensional objects according to the presentist view) must flow *equally*. This means that if the twins were three-dimensional objects there would be no time difference when A and B directly compare their proper times at M. Therefore *the twin paradox effect, if formulated in terms of the presentist view, would be impossible*. This shows the incorrectness of our initial assumption – that A and B exist only as three-dimensional objects which are subjected to an objective flow of time as required by the relativized version of presentism. *The fact that the twin paradox effect and the experiments that confirm it would not be possible if the twins were three-dimensional objects completely rules out the relativized version of presentism.*

Let us state it again: the twin paradox is consistently explained if A's and B's worldtubes are real four-dimensional objects; then twin A exists at all events comprising his worldtube. As indicated above the time difference of five years when the twins 'meet' at M comes from the different lengths of the twins' worldtubes between the events D and M; that is, the different amounts of proper times of the twins between D and M. The four-dimensionalist view offers a natural explanation of why the acceleration does not affect the amount of proper time measured by twin B in Fig. 5.11: the acceleration which twin B suffers is merely an indication that his worldtube is curved, but this curvature does not affect his proper time since *the length of a worldtube does not change if it is curved*. The only role of the acceleration in Fig. 5.11 is to show that B's worldtube is curved and that it is a different path from event D to event M which due to the pseudo-Euclidean nature of Minkowski spacetime is shorter than the path along A's worldtube. If B's worldtube is straightened and superimposed on A's worldtube, the length of B's worldtube will be equal to the segment DI of A's worldtube.

Again, the spacetime explanation of the twin paradox effect shows that there is nothing mysterious in it. It is simply the triangle inequality in the pseudo-Euclidean geometry of Minkowski spacetime.

The last argument, which I will discuss here, against presentism and relativization of existence involves accelerated observers in relativity [33, Chap. 5]. Consider an accelerated observer in Minkowski spacetime whose worldtube is shown in Fig. 5.12. The figure also shows the worldtube of a digital clock. Due to the fact that the observer's worldtube is curved (i.e. because the observer accelerates), the presents (i.e. the instantaneous spaces) which correspond to the events P and Q of his proper time are not parallel and intersect at event O. If we assume that existence is relativized, the world would be three-dimensional and all objects would be also the familiar three-dimensional bodies. Therefore both the accelerating observer's and the digital clock's worldtubes would not be real. At event P the clocks would exist at its 11th second. The clock before its 11th second would be in the *past* of the accelerated observer at event P. However, the clock after its 0th second lies in the *future* of the accelerated observer at the *later* event Q. So what is past at P turns out to be future at Q. We arrived at such nonsense because we assumed that the world was three-dimensional. Both presentism and relativized presentism lead to this nonsense since they both regard reality as an evolving three-dimensional world; the recently revived growing (or evolving) block universe view leads to the same absurdity (see below).

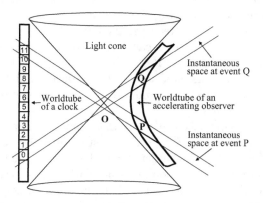

Figure 5.12: Presentism, relativised presentism, and the growing (or evolving) block universe view all lead to nonsensical results when adopted by an accelerating observer. As the worldtube of such an observer is curved the presents (i.e. the instantaneous spaces) that correspond to two moments of his proper time intersect at event O. All readings before the 11th second of the digital clocks are *past* for the accelerated observer when event P is 'now'. However, the *same* events lie in the observer's *future* when event Q is 'now'

Therefore, accelerated observers in special relativity cannot hold the presentist, i.e. the three-dimensionalist world view (pre-relativistic or relativized) since it leads to complete nonsense. In this sense, the very existence of accelerated observers in special relativity is another manifestation of the four-dimensionality of the world (since spacetime is not divided into past, present, and future events and no such nonsense arises).

I would like to state it again, that the failure of the presentist view, when applied to accelerated observers in relativity, is another argument against relativization of existence because the relativized version of presentism regards the world of an observer as three-dimensional. Therefore, not only would the twin paradox effect (which is an absolute relativistic effect with no relativity of simultaneity involved) be impossible if the world were three-dimensional, but relativity of simultaneity and its two specific manifestations – length contraction and time dilation – would be also impossible in a three-dimensional world (since relativized existence is not an option).

A physicist might disagree with the above argument against the three-dimensionality of the world by pointing out that there are constrains on the size of an accelerated frame in special relativity [54, p. 168]. These constrains result from the fact that a *global* coordinate system associated with the accelerated observer cannot be extended to the left in Fig. 5.12 beyond event O since coordinate time makes no sense beyond that event – what is past time at event P is future time at the later event Q. Also, the coordinates of the events located in the spacetime region between the past and the future light cones, which contains the digital clock, cannot be determined by sending and receiving light signals [54, p. 168]. The event O (in fact it is a two-dimensional surface) acts as a horizon for the accelerated observer – no signals can be received from and sent to the spacetime region lying to the left of that event.

All this is, of course, correct but these objections are concerned *only* with the *description* of the events of spacetime and have nothing to do with the question of whether or not they exist. The fact that an accelerated observer cannot communicate with events located to the left of event O (between the past and the future light cones at O) is irrelevant for the existence of these events. This is best demonstrated by taking into account that all instantaneously comoving inertial observers at the different events of the worldtube of the accelerated observer can communicate with the events to the left of event O (between the past and the future light cones at O). And if these events exist for the comoving inertial observers they should exist for the accelerated observer as well.

As mentioned in Chapter 1 several well-known physicists [7]-[9] have recently defended the key feature of presentism (the reality of time flow). They considered versions of the so-called growing or evolving block universe according to which the past and the present exist, but the future does not. On that view the world is not a "completed" block universe, but a growing one. However, this view assumes the existence of a *privileged* hyper-surface – the present at which the future events come into existence. Such a privileged hyper-surface contradicts relativity and also leads to the same nonsensical results shown in Fig. 5.12 (for a more detailed refutation of this recent attempt to deny the reality of spacetime see [33, Chap. 6]).

6 EINSTEIN AGAINST THE WHOLE WORLD IN 1916 – IT IS AN ILLUSION THAT GRAVITY IS A FORCE, GRAVITY IS CURVATURE OF SPACETIME

We are all enormously lucky that there is no democracy in science. Otherwise, such an ingenious theory as general relativity would have never been discovered. Everyone in the world not only believed but experienced through their weight that gravity is a force. In 1916 when Einstein published his general relativity he showed that gravity was not a force, but a manifestation of the curvature of spacetime. In 1916 he was alone against the whole world. And the ultimate judge – the experimental evidence – ruled in his favor.

As Parmenides [20, p. 64] wrote about his theory of being that "there are very many signs" of it, almost the same could be said about the discovery that gravity is not a force. There existed signs that could have ultimately led to a new theory of gravitation – e.g. Galileo's observation that bodies of different weight fall at the same rate, and at least two problems with Newton's notion of gravitational force [51, Chap. 3]. A rigorous conceptual analysis of Newton's gravitational theory could have revealed them, long before Einstein. The first of those problems was realized by Einstein most probably in November 1907 and this insight set him on the path toward his theory of general relativity (quoted from [34]):

> I was sitting in a chair in the patent office at Bern when all of a sudden a thought occurred to me: "If a person falls freely he will not feel his own weight." I was startled. This

simple thought made a deep impression on me. It impelled
me toward a theory of gravitation.

Einstein was so impressed by this insight that he called it the "happiest thought" of his life [34]. Then Einstein needed eight years to arrive at his general relativity. This is an extremely short period for such
a revolutionary theory. However, such a thinker like Einstein could
have discovered the new theory of gravitation much sooner. I think he
was delayed by the unproductive doctrine of relativism he shared at
the time he created special relativity and later when he struggled with
the theory of gravity. That doctrine is responsible for the inadequate
names[1] of his two theories of *relativity* – there is nothing fundamen-
tally relative either in special relativity or in general relativity (all
relative quantities in both theories are possible *only* due the existence
of the *absolute* spacetime). Perhaps relativism had been behind Einstein's initial hostile reaction at Minkowski's four-dimensional physics.
Sommerfeld's recollection of what Einstein said on one occasion provides an indication of his initial attitude toward Minkowski's work:
"Since the mathematicians have invaded the relativity theory, I do not
understand it myself any more" [52].

However, later Einstein adopted Minkowski's view of spacetime
and his four-dimensional physics and quickly completed (in 1915) perhaps the deepest physical theory of all time – his general relativity
which regards gravity not as a force but as a manifestation of the non-
Euclidean geometry of spacetime (the curvature of spacetime is caused
by the presence of massive bodies).

Here I would like to demonstrate what a research strategy, based
on the reliable knowledge that the world is four-dimensional and that
four-dimensional physics is in fact spacetime geometry, might have
produced, if Einstein used it together with his happiest thought. Let
us imagine that Einstein examined his thought – a falling body that
does not feel the force of gravity – and realized that *a falling body
does not resist its fall* since a body resists its motion only if it is
subject to a force (recall that by Newton's second law a force is only
needed to overcome the resistance the body offers to its acceleration).
Such thought, which follows unavoidably from Einstein's "happiest
thought," might have paused Einstein a bit since its meaning is that
a falling particle moves by inertia. But how can a falling body move

---

[1] In his "To Albert Einstein's Seventieth Birthday" [52, p. 99] Sommerfeld pub-
licly expressed the widespread uneasiness about the names of Einstein's two theories – "the widely misunderstood and not very fortunate name of "theory of relativity"."

by inertia, if it accelerates? Ironically, Einstein's relativism might have again helped him, this time to regard acceleration as relative (as he actually did it). Imagine also that Einstein carefully analyzed Minkowski's view of reality as an absolute four-dimensional world and arrived at the results discussed in Chapter 5.

Then the path to the idea that gravitational phenomena are manifestations of the curvature of spacetime would have been open – the experimental fact that a falling body accelerates (which means that its worldtube is curved), but offers no resistance to its acceleration (which means that its worldtube is not deformed) can be explained only if the worldtube of a falling body is *both curved and not deformed*, which is impossible in the flat Minkowski spacetime where a curved worldtube is always deformed. Such a worldtube can exist only in a non-Euclidean spacetime whose "straight" worldlines (called geodesics) are naturally curved due to the spacetime curvature, but are not deformed. This insight provides an unexpected answer to the above question (How can a falling body move by inertia, if it accelerates?) – as there are no straight and parallel worldlines in curved spacetime, the geodesic worldlines of two bodies moving by inertia will either converge or diverge (called geodesic deviation) and as a result it will *appear* that the bodies accelerate toward or away from each other).

Figure 6.1: Gravitational "attraction" in two dimensional spacetime

To see better how a worldline can be curved but not deformed consider the meridians and the equator on a globe as shown in Fig. 6.1. As there are no straight lines on the curved surface of a sphere, the analog of straight lines are the geodesics (the great circles obtained by the intersection of the sphere and plane surface passing through the sphere's center), i.e. the meridians and the equator in the case of the globe. They are naturally curved due to the curvature of the globe's surface, but are not deformed. If we try to curve a meridian further, we deform it.

Einstein would have realized that Minkowski's program "the laws of physics can find their most complete expression as interrelations between these worldlines" [12, p. 112] had produced another stunning result – gravitational phenomena are not caused by a force, but are merely effects due to the curvature of spacetime.

To see that gravitational phenomena are indeed geometrical effects on a curved spacetime, consider two balls on the equator as shown in Fig. 6.1, which start to move upward in directions that are perpendicular to the equator. As the balls recede from the equator they start to approach each other. If we are unaware that the balls move on the surface of a sphere and implicitly assume that they move on a plane, we will explain the shortening of the distance between them as caused by a force of attraction between the balls. This is how Newton explained the gravitational phenomena (attraction between bodies). However, when we take into account the real situation – that the balls are on the curved surface of a sphere – it is self-evident that no force is acting on them since the balls move by inertia on the curved surface; the reason they approach each other is that there are no parallel lines on the surface of the sphere.

In fact, this explanation is a bit simplified since it considered the surface of an ordinary globe and regarded the two meridians, along which the balls moved, as the *trajectories*[2] of the balls. The real situation depicted in Fig. 6.1 represent a two dimensional curved spacetime and the meridians are the *worldlines* of the balls. The two worldlines converge which in the ordinary three-dimensional language means that the balls are approaching each other.

Let us now return to the falling body. Such a body is represented by a geodesic (i.e. not deformed) worldtube, which means that the body moves by inertia, i.e. without resisting its fall. This fact is captured in the geodesic hypothesis in general relativity, which states that the worldline of a free particle is a timelike *geodesic* in spacetime. This hypothesis[3] is "a natural generalization of Newton's first law" [65], that is, "a mere extension of Galileo's law of inertia to curved spacetime" [66]. This means that *in general relativity a particle, whose worldline is geodesic, is a free particle which moves by inertia.*

---

[2] A trajectory of a moving body is the projection of its worldline on space.

[3] The fact that a falling body does not resist its fall, which means that it moves by inertia, is called hypothesis by tradition; this fact is experimentally confirmed – a falling accelerometer, for example, reads zero resistance since it measures acceleration through resistance.

When a falling body reaches the ground it is prevented from falling (i.e. from moving by inertia) and *resists* that change by exerting an *inertial* force on the ground, which has been traditionally called gravitational force. That the force with which the falling body resists the change in its inertial motion is inertial, is nicely demonstrated when we consider the worldtube of the body when it hits the ground. The worldtube of the body (when it is on the Earth's surface) is *deformed*, exactly like the worldtube of an accelerating body is deformed, and therefore the restoring static force that is caused by the deformation is inertial like the restoring static force originating from the deformed worldtube of the accelerating body. The mass of the body, when at rest on the ground, has been traditionally called passive gravitational mass, but we see that it is inertial mass – it is the measure of the resistance the body offers when prevented from moving by inertia while falling.

As we see it turns out that the inexplicable equivalence of the inertial mass and the passive gravitational mass (and the equivalence of the inertial and gravitational forces), which Einstein merely postulated as the equivalence principle, found a natural explanation: inertial and gravitational masses and forces are equivalent since they are the same thing – they are all inertial. Minkowski's program regarding physics as spacetime geometry again helped us to find another deep explanation – that of the equivalence principle.

The example with the falling body also nicely explains the status of acceleration in general relativity (i.e. in curved spacetime). A falling body accelerates (while moving by inertia) but that acceleration is *apparent* since it is caused by the curvature of spacetime induced by the Earth's mass – the geodesic worldline of the Earth's center and the body's geodesic worldtube converge toward each other (there are no straight and parallel worldlines in curved spacetime) which we perceive as a falling body. The apparent acceleration, caused by the geodesic deviation of the body's geodesic worldtube and the geodesic worldline of the Earth's center, is relative and involves no deformation of the bodies worldtube. By contrast, the worldtube of the body, when it is at rest on the Earth's surface, is *deformed* which indicates that the body's acceleration is absolute. Indeed, the body resists its being prevented from moving by inertia while falling by exerting a real inertial force on the ground.

The status of acceleration in flat and curved spacetime is the same – in both cases the acceleration of a body is *absolute* if its worldtube is *deformed*. In curved spacetime there is a second acceleration – apparent (relative) acceleration (caused by geodesic deviation) – which is a manifestation of the spacetime curvature. All gravitational accelerations (of falling bodies and of planets orbiting the Sun) are such apparent accelerations through which the non-Euclidean geometry of spacetime manifests itself.

The example with the falling body makes it possible to clarify another misunderstanding ultimately coming from Mach – that the shape a geodesic worldtube is determined by all masses in the Universe and therefore when a body is deviated from its geodesic path the inertial force with which the body resists the change in its inertial motion is determined by all the masses. That all masses in the Universe determine the global curvature of the Universe is undeniable, but the claim that the inertial force arising in the deformed worldtube of a body is simply untrue. The shape of the geodesic worldtube of a body falling toward the Earth is determined *predominantly* by the spacetime curvature induced by the Earth's mass. There are small contributions from the Moon and the Sun and practically zero contributions from the distant masses in the Universe. When the body is at rest on the Earth's surface its worldtube is deformed and the resulting inertial force (traditionally called gravitational force or the body's weight) originates from the four-dimensional stress in the body's own worldtube; that is, *the body's inertia originates in the body itself.* As the shape of the falling body's geodesic worldtube is almost entirely determined by the spacetime curvature caused by the Earth's mass, the question of whether the Earth's mass has a contribution in the body's inertial force (exerted on the ground) will remain open until we understand how the Earth curves the spacetime around its worldtube.

Had Einstein followed Minkowski's program that four-dimensional physics is nothing more than spacetime geometry, he would have arrived at the conclusion that *gravitational phenomena are not caused by gravitational interaction* since they are mere manifestation of the geometry of curved spacetime. Then the conclusion that gravitation is not a physical interaction is inevitable. For example, the planets are free bodies which move by inertia (since their worldlines are geodesics) and as such they do not interact in any way with the Sun because *inertial motion does not imply any interaction.* A complete understanding of gravitational phenomena requires one more step – not to try to quantize gravity (since there is no gravitational interaction to quantize), but to resolve the major open question of how matter curves spacetime.

However Einstein followed a different path and even such a great thinker was unable to free himself from the deceivingly overwhelming evidence that gravity is interaction and that gravitational energy is involved in the gravitational phenomena. See [51, App. C] for a discussion on the existence of gravitational energy in general relativity and on a possible reason for the unsuccessful attempts to create a theory of quantum gravity – that gravity is not a physical interaction – which does not appear to have been examined so far.

7 ANOTHER STRANGE WORLD – THE WORLD OF QUANTA

> *Continuous* existence in time of quantum
> objects may turn out to be another illusion.

After the advent of relativity and quantum mechanics it has been often thought of a serious problem in modern physics – the apparent contradiction between the deterministic theory of relativity and the probabilistic quantum theory. Sometimes this view has been used by people who do not feel comfortable with the spacetime world view and who caution that the implications of relativity should not be taken too seriously because it clearly contradicts the experimentally-confirmed feature of quantum mechanics – quantum randomness. One thing, however, is crystal clear – as experiments do not contradict one another, the quantum experiments, which would be impossible if quantum randomness did not exist, cannot contradict the relativistic experiments, which would be impossible if spacetime did not exist.

I think three issues should be kept in mind since they demonstrate that the apparent contradiction between relativity and quantum physics is exactly that – apparent:

(i) Although special relativity is not fully applicable at the quantum scale (since its equations of motion manifestly fail to describe the behaviour of quantum objects), it is an undeniable fact that spacetime is still the arena at that scale.

(ii) Quantum randomness exists at the microscopic (not at the macroscopic) scale. No one has ever seen or heard of a train going simultaneously through two tunnels!

(iii) The only thing quantum randomness tells us is that quantum objects are not worldlines in spacetime and nothing more; it is totally irrelevant to the question of the reality of spacetime.

In Chapter 4 we saw why the emerging quantum physics helped Einstein to discover special relativity. Before Planck's ground-breaking discovery in 1900 that light is emitted as packages of electromagnetic energy (quanta) it was believed (due to Young's double-slit experiment and Maxwell's electrodynamics) that light was an electromagnetic wave, which is emitted *continuously*, not in quanta. But as a wave is a disturbance in a medium, the fact that light propagates from the Sun to the Earth, for example, implied that space is some kind of a medium, which was called luminiferous ether. So it appeared that Young's experiment, which proved that light was a wave, indirectly proved the existence of the ether. But if the ether did exist, the Earth's motion in it should be detectable, which would put pressure on Galileo's principle of relativity according to which, in this case, the Earth's motion cannot be discovered with mechanical experiments. That pressure was lifted when the Michelson and Morley experiment failed to detect the Earth's motion in the ether, but the puzzle remained – if the ether existed, why motion with respect to it cannot be discovered.

Einstein solved the puzzle by merely postulating that the ether did not exist. He was able to do this since he had some answer to the obvious question – how would light even exist if the ether did not exist (because, as a wave, light is a *disturbance* of the ether). Unlike Lorentz and Poincaré, Einstein immediately took Planck's hypothesis of light quanta probably more seriously than Planck himself and even developed it further by showing that not only is light emitted as quanta (as Planck conjectured), but it is also absorbed as quanta. Einstein employed this development of Planck's original idea to *explain* the photo-electric effect for which he received the Nobel Prize in Physics in 1921. Einstein's work on the emerging concept of light quanta enabled him to realize that the newly discovered *particle* aspect of light implies that light is not necessarily a disturbance of a medium and therefore does not necessarily require the existence of the ether for its own existence.

The particle aspect of light was virtually forced by experimental results upon the physicists at the turn of the 19th and 20th centuries. Neither the emission of light by a black body nor the photo-electric effect could be explained by the wave theory of light. Planck's original insight that light should be emitted as quanta explained the first experimental result (the ultraviolet catastrophe), whereas Einstein's development of the idea – that light should be also absorbed as quanta

– explained the photo-electric effect.

In the beginning of the 20th century the emerging quantum theory was no less paradoxical then the theory of relativity. Young's double-slit experiment proved that light was a wave. The experiments of emission of light by a black body and of the photo-electric effect proved that light behaved as a particle – a quantum or a photon (as it was later called). As experiments do not contradict one another, since then the tantalizing question has been "How can light be *both* a wave and a particle?" Einstein believed that it was more appropriate to say that light was something third – *neither a wave nor a particle.*

As quantum theory gives us only probabilities of experimental outcomes it is often stated that it tells us nothing about the measurement-independent nature of the quantum objects involved in the experiments and therefore the question "What is the quantum object *by itself?*" is incorrect or metaphysical at best. Such a position is rather paradoxical because at the same time particle physics is studying the properties of quantum objects and trying to discover new 'particles' predicted by the Standard Model (most recently – the famous Higgs boson). I think everyone would agree that we will never gain deep understanding of the world if legitimate physical questions – such as those of the nature of the entities living at the quantum scale – are ignored by labelling them metaphysical or philosophical.

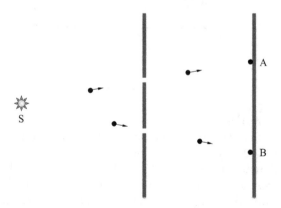

Figure 7.1: Double-slit experiment with particles

What is ultimately responsible for all apparent quantum paradoxes is the wave-particle duality of quantum objects. Feynman notably acknowledged the weirdness of quantum phenomena – "I can safely say that nobody understands quantum mechanics" [57, p. 129]. From time to time Feynman's admission is downplayed by saying that he was joking. Ironically, that could have been precisely the case – Feynman might have really joked, especially given the fact that one can only joke about a problem so intractable that all attempts to find a solution have failed so far.

To demonstrate that indeed nobody understands quantum mechanics let us consider the same interference experiment – Young's double-slit experiment – which Feynman discussed and "which has been designed to contain all of the mystery of quantum mechanics" [57, p. 130]. As shown in Fig. 7.1 this experiment uses a source S, a plate with two slits, and a screen. Let us first assume that the source emits ordinary particles (e.g. very small balls). The most particles will be stopped by the plate and will not reach the screen. Those particles which go through the slits will hit the screen at the two locations A and B which are situated behind the slit.

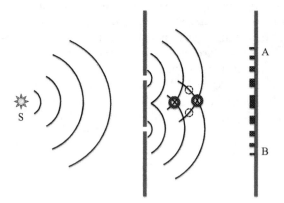

Figure 7.2: Double-slit experiment with waves

Consider now Fig. 7.2 which shows what happens when the source S emits waves, i.e. disturbances in the medium surrounding the source, the plate with the two slits and the screen. When the emitted disturbance (wave) reaches the plate it will be also stopped there. The disturbance can be transmitted to the other side of the plate only through the slits. So it appears that the two slits emit two secondary waves (disturbances) resulting from the first disturbance. Then the two secondary disturbances start to interact. When the maxima of the two waves (depicted in Fig. 7.2 by the arcs) coincide (i.e. when the arcs intersect) the resulting disturbance is greater. This kind of interaction is called constructive interference (two examples of it are marked by the black-and-white rings in the figure). When a maximum of one of the waves coincides with a minimum (the region between two arcs) of the other wave the two disturbances cancel out. This kind of interaction is called destructive interference (two examples of it are marked by the dashed circles in the figure).

The result of an experiment with light shows an interference pattern on the screen – a sequence of bright (constructive interference) and dark (destructive interference) fringes. The brightest fringe is located in the middle of the line AB. It should be particularly stressed that *the interference pattern is present only if light passes through both slits*. If one of the slits is closed, there is no interference pattern.

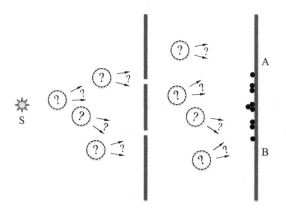

Figure 7.3: Double-slit experiment with a beam of electrons

When electrons were used in the double-slit experiment the result was amazing – as shown in Fig. 7.3 an interference pattern was observed. This meant that the electrons behaved like a wave. There had been suggestions that the interactions of the negatively charged electrons, which passed through the slits, produced the interference pattern. This was an attempt to avoid the mind-blowing implication that the interference pattern implies that every electron passed through both slits.

To test this classical explanation of the interference pattern, the experiment was repeated with *single* electrons (one electron at a time) as shown in Fig. 7.4. After some time the same interference pattern was formed on the screen. The mind-blowing implication was proved – *every electron goes through both slits*.

Real double-slit experiments have been performed with *single* electrons and photons and an interference pattern has always been observed [58]-[61]. This can be only explained by the wave-like behaviour of *individual* electrons and photons – every *single electron and photon goes through both slits at the same time and then interferes with it-*

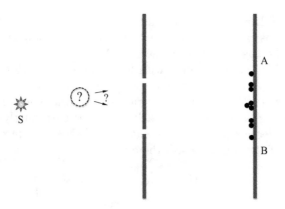

Figure 7.4: Double-slit experiment with single electrons

self before hitting the screen behind the slits at a specific *location*. It should be noted that the double-slit experiments with single electrons and photons, in fact, proved both their wave and particle behaviour – the wave behaviour was proved by the presence of an interference pattern, whereas the particle behaviour was proved by the fact that every electron and photon hit the screen at a specific location and are therefore registered as localized entities.

The mystery is so disturbingly obvious – how can a photon or an electron, which is registered as a *localized* entity (and only where the bright fringes of the interference pattern are formed) go through *both* slits? Feynman's advice is [57, p. 129]:

> Do not keep saying to yourself, if you can possibly avoid it, 'But how can it be like that?' because you will get 'down the drain', into a blind alley from which nobody has yet escaped. Nobody knows how it can be like that.

Feynman appears to have given up the hope that we will be able to solve the mystery and understand the strange world at the quantum scale. But to help anyone who studied the quantum phenomena he suggested that a working strategy would be to merely accept their absurdity [62]:

> The theory of quantum electrodynamics describes Nature as absurd from the point of view of common sense. And it agrees fully with experiment. So I hope you can accept Nature as She is – absurd.

Feynman's advice might be a good temporary strategy to continue the quest for understanding of the quantum world, but no real advancement would be possible without a genuine understanding of the intimate mechanism of quantum phenomena. Another good strategy is to carry out rigorous conceptual analyses of the existing theoretical and experimental evidence, one of whose aims is to identify some implicit assumptions, which might have been leading to apparent paradoxes. Although there have been some signs of a doubtful tendency in fundamental physics in recent decades – that such analyses are rather old-fashioned – the history of physics (especially the contributions of Galileo, Newton, Einstein, and the founders of quantum physics) has convincingly demonstrated that conceptual analyses are physics at its best.

Let us see how a brief conceptual analysis of the interference experiments with single photons (or electrons) can help us identify what appears to be an important implicit assumption, which may hold the key for a breakthrough in our true understanding of quantum phenomena. Let me start with Dirac's brilliant conceptual analysis of interference of photons, which helped him arrive at the conclusion that every photon of a photon beam should participate in both components of the split beam and should *interfere only with itself* [63]:

> Suppose we have a beam of light consisting of a large number of photons split up into two components of equal intensity. On the assumption that the intensity of a beam is connected with the probable number of photons in it, we should have half the total number of photons going into each component. If the two components are now made to interfere, we should require a photon in one component to be able to interfere with one in the other. Sometimes these two photons would have to annihilate one another and other times they would have to produce four photons. This would contradict the conservation of energy. The new theory, which connects the wave function with probabilities for one photon, gets over the difficulty by making each photon go partly into each of the two components. Each photon then interferes only with itself. Interference between two different photons never occurs.

Although there have been attempts to question Dirac's conclusion – that a photon can interfere only with itself – there exists undeniable experimental evidence supporting that mystery. Even the evidence from our daily experience is overwhelming – no mobile telephone and no radio and TV broadcast would be possible if *different* photons did interfere. We, as species, would not have survived if Dirac's analysis and conclusion were wrong because in such a case photons reflected from different objects would interfere before hitting our eyes' retina and we would be unable to see the world properly.

So the mystery that every photon and electron in the interference experiments with single photons and electrons passes through both slits and interferes only with itself is an experimental fact. The question is whether it is really a mystery which we should merely accept and give up any attempts to solve it. Let us start with what is undeniable – every electron, for example, (i) goes through both slits, and (ii) is registered as a localized entity. What appears to cause the mystery is the belief that such a localized entity somehow goes through both slits. However, that belief is based on the implicit assumption that if the electron hits the screen behind the slits as a localized entity, it was such an entity at every moment of time before that as well. More specifically, the implicit assumption is that an electron exists as a localized entity at *all* moments of time, i.e. that an electron exists *continuously* in time as a localized entity. In other words, the mystery is caused by the belief that an electron is after all a particle (since it *appears* to be registered as such), which continuously exists in time, but somehow passes through the two slits.

If we replace the identified implicit assumption (continuous existence in time) with its (now obvious) alternative – that an electron exists *discontinuously* in time as a localized object – the mystery disappears at once since the appearing and disappearing constituents of the *same* electron can go through all slits at their disposal and then reunite (the electron interferes with *itself*), but always hits the screen as localized (particle-like) entity (when the first constituent of the electron collides with the screen it is trapped there and due to the jump of the boundary conditions all subsequent constituents appear and disappear at the same location).

The idea of bringing the concept of atomism (more specifically of discreteness) to its logical completion – discreteness not only in space, but in time as well (4-atomism) – was proposed in the eighties (see [64]) but unfortunately remained unnoticed. I think it is worth examining this radical idea by testing its predictions [64] and eventually deducing new ones. Its careful examination is especially warranted by the fact that the assumption that electrons (and quantum objects in general) do not exist continuously in time appears to provide unexpected but reasonable conceptual answers to probably all quantum puzzles. Here, in addition to the double-slit-type experiments, I will list only three

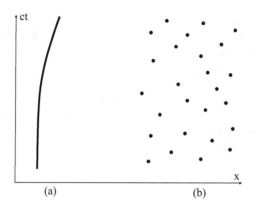

Figure 7.5: (a) A classical particle exists continuosly in time and is represented by a timelike worldline in spacetime. (b) If a quantum object (e.g. an electron) exists discontinuously in time it is represented by a different spacetime structure – a probabilistic distribution of entities localized in spacetime (the "constituents" of the "disintegrated" worldline of the classical electron), which is forever given in spacetime.

examples (for a little more detailed conceptual account of the idea that quantum objects may exist discontinuously in time see [33, Chap. 10]):

- The mysterious Compton frequency of the electron given by the expression $f_C = m_e c^2 / h \approx 10^{20} s^{-1}$ can be interpreted as the frequency of appearance / disappearance of the electron's constituents.

- Superposition, quantum jumps, and tunnelling can be consistently explained by the assumption that what was regarded as a spatially indivisible electron might in fact have a structure in time (rather in spacetime).

- The idea of discontinuously existing quantum objects also provides a surprisingly coherent explanation of what has been regarded as a serious problem of modern physics – the mentioned in the beginning of the chapter apparent contradiction between the deterministic theory of relativity and the probabilistic quantum phenomena. That apparent contradiction presupposes that quantum objects are worldlines in spacetime. But *worldlines are not the only thinkable spacetime structures.* If, for the sake of argument, the quantum object (an electron, for example) is a different spacetime structure – atomistic in spacetime – then the points of the electron's "disintegrated" worldline will be randomly scattered all over the spacetime region where the electron wavefunction is different from zero (Fig. 7.5). This explains perfectly the probabilistic features of the electron in spacetime: the behaviour of an electron can be described only in terms of probability, but on the other hand the whole history of the electron in time (the probabilistic distribution of all of its "constituents," representing the electron at all moments of its history) is *forever given* in spacetime as a "frozen" probabilistic distribution. Had Minkowski lived longer he might have expressed this situation with the magical phrase "predetermined probability."

8 SPACETIME AND THE ILLUSION THAT TIME FLOWS

> Don't confuse language with reality. Human language is far better at capturing human experience than at expressing deep physical laws.
>
> B. Greene [70]

In Chapter 4 we saw what intellectual power Einstein needed to identify one of the greatest illusions – that time was absolute, i.e. that there existed an absolute 'now', at which the whole world existed, which meant that simultaneity was absolute because every 'now' was arriving simultaneously in the whole universe. As the theory of relativity showed that there is no absolute time (and therefore no absolute division of events into past, present, and future), it follows that there should be no absolute flow of time either. But the meaning of this conclusion had not been immediately realized after the advent of special relativity. Initially, Einstein assumed that time was *relative*, i.e., that the flow of time was different for different observers in relative motion. But as we saw in Chapter 5 Minkowski powerfully demonstrated that relative times are manifestation of an absolute reality – the four-dimensional spacetime (recall – relative times would be impossible, if spacetime did not exist). As in the Minkowski world, all events are *equally* existent, and therefore are not objectively divided into past, present, and future, there is no flow of time. And, indeed, no flow of time is possible in spacetime (in a four-dimensional world) since the very essence of time flow is that only *one* moment of time exists, whereas *all* moments of time exist in spacetime. Therefore, Minkowski's spacetime formulation of Einstein's special relativity clarified the meaning of the implication of special relativity that there is no absolute flow of time – there is no time flow at all.

However, it is an *inter-subjective*[1] fact that every human realizes one single moment of time that constantly changes. What causes this experience? Does our feeling that time flows reflect an objective fact? It certainly does not reflect a fact that only one event of the worldtube of a person exists since the entire worldtube is a real four-dimensional object as we saw in Chapter 5. So the fact that we realize ourselves at one moment should reflect some other feature of reality. Trying to answer this type of questions and, most importantly, to reconcile two facts – the reality of worldtubes (and the reality of spacetime) and the inter-subjective fact that we realize ourselves at one single moment of time – Weyl proposed what appears to be the only meaningful concept of time flow in a four-dimensional world [71]:

> The objective world simply *is*, it does not *happen*. Only to the gaze of my consciousness, crawling upward along the life line of my body, does a certain section of this world come to life as a fleeting image in space which continuously changes in time.

For the first time in the history of science the concept of consciousness was needed to reconcile the interpretation of a physical theory with the undeniable fact that we all feel the flow of time. The interpretation of the original three-dimensional formulation of special relativity given by Einstein in 1905 did not need the concept of consciousness since it seemed to be consistent with the belief that time flows. However, after Minkowski's four-dimensional formulation of special relativity it became clear that our feeling of time flow does not reflect a *physical* fact since all events exist in spacetime which means that the events are not objectively divided into past, present, and future. It is possible to object that the spacetime world view does not require the concept of consciousness, since we can obtain all results of special relativity without it. Indeed, relativity *as a physical theory* does not need that concept. But when we want to understand how we can have the feeling that time flows in the Minkowski four-dimensional world, it appears that Weyl's proposal holds the greatest promise for the resolution of the apparently insurmountable contradiction between the physical theory of relativity (and the experiments that support it) and our everyday experience. Moreover, so far, no one has found a

---

[1] I prefer to call this fact inter-subjective, not objective, since we cannot apply the scientific method properly and to verify *experimentally* the existence of this fact. However, we all have the same experience and for this reason we regard this experience as something more than other subjective experiences.

way, which does not involve our consciousness, to reconcile the space-time world view and the fact that whatever we perceive happens only at the present moment.

As we saw in Chapter 2 in order to resolve logical contradictions arising from the assumption that only the present moment of time existed, Aristotle and Augustine conjectured that the division of time into past, present, and future (i.e., the flow of time) belonged only to the mind. After Minkowski's revolutionary ideas, Weyl was led by the interpretation of a physical theory (special relativity which deals with the physics of spacetime) to the same conclusion.

Weyl almost explicitly regarded consciousness as an entity which is 'moving' along the worldtubes of our bodies and which makes us aware of ourselves and the external world at one event. He did not pretend to understand what the consciousness is; he simply assumed that *something* was making us self-conscious and postulated that that something, which he called consciousness, moves along the worldtubes of our bodies and creates the feeling that time flows.

Let me stress that I do not see any other way to explain that feeling in the four-dimensional world we inhabit. You will also understand why Weyl so bravely went ahead with his bold idea that the flow of time is mind-dependent, if you (i) keep in mind the reason which forced Aristotle and Augustine to conjecture that there existed a link between time and the mind (and the reasons that made the Eleatics to argue about essentially the same), and (ii) try to explore Weyl's idea in more detail. To do the latter, assume (even if you do not believe it) that the world is four-dimensional and it is your consciousness that creates the feeling of time flow. According to Weyl, your consciousness crawls upward along the worldtube of your body and reads the information from your senses that is stored in your brain, but incorrectly interprets this information in a sense that a three-dimensional world exists and is constantly changing in time. As a result of this incorrect interpretation, you would be completely convinced that you were living in a three-dimensional world which is evolving in time.

His idea that the flow of time is not objective but is mind-dependent has been frequently quoted, but no one has examined it more rigorously. In order to see the need for such an examination, let us briefly discuss the implications of Weyl's idea.

At first glance this idea appears to contradict the spacetime world view since Weyl assumed that the consciousness *moves* in spacetime where no motion is possible. This contradiction needs careful scrutiny to determine whether it is apparent or real because of the importance of Weyl's conjecture as the only candidate to explain our feeling of time flow in the four-dimensional world. It is obvious that there will be an immediate contradiction only if it is assumed that the consciousness is a *physical* entity in spacetime. Although no one has detected a consciousness wandering in the brain, let alone in the street, many would probably regard the assumption that the consciousness is not physical as not belonging to science. I think such an approach of label placing itself is deeply unscientific. When the stakes are at the highest possible level – our view of reality – nothing should be taken for granted. All, even purely logical, options should be on the research table. We have nothing to lose if we consider all possible ideas, but we may have a lot to lose if we do not do it.

At present, neuroscience assumes that consciousness is a result of brain processes, which are ultimately governed by the laws of physics. On this view, the consciousness is part of the physical world, whose basic arena is spacetime, and therefore the consciousness cannot move in spacetime. In this situation we have no choice but to question either the spacetime world view or today's understanding of the consciousness. On the one hand, the arguments supporting Minkowski's view are, as we have seen, overwhelming. On the other hand, there is a consensus among scientists that neuroscience does not fully understand consciousness. Some researchers even doubt that a science of consciousness is possible since they think "that 'consciousness' is not a viable target of scientific research" [72].

If consciousness is not physical we essentially adopt the old dualistic interpretation of consciousness and matter. As according to this dualistic interpretation of consciousness, it is not a physical entity, it does not obey the relativistic and any other physical laws. It is interesting that the usual arguments against the philosophical dualistic view do not hold in this interpretation of Weyl's conjecture. According to those arguments, if consciousness or the mind were not physical, how could the body affect the mind and how could the mind tell the body what to do? On Weyl's view there is no interaction between the consciousness and the body's worldtube. The consciousness simply "reads" the information from our senses stored in the brain and this is the only "interaction" in which the consciousness is involved. The body's worldtube does not influence the consciousness at all, and the consciousness does not affect the worldtube in any way.

As we have been taking it as exceedingly obvious that it is the consciousness that calls the shots and tells our body what to do, the passive role of the consciousness in Weyl's explanation (only to "read" the information from our senses stored in the brain) may seem outrageously wrong to many. If you tend to share such disagreement, I think it will be wise to consider these two points first:

- I believe no one doubts that Weyl was fully aware of the taken for granted role of consciousness – to tell the body what to do. The very fact that he went public with his bold idea reveals that he was forced to do it by the impossibility to refute either (i) the four-dimensionality of the world (proved by the relativistic experimental evidence) or (ii) the inter-subjective fact that we realize ourselves only at the moment 'now' which constantly changes. It appears plausible that Weyl discovered the only logically possible way to reconcile these two facts by resorting to the concept of consciousness and assigning a very passive role to it. Moreover, we have seen in Chapter 2 how the Eleatics anticipated, Aristotle suspected, and Augustine conjectured that the flow of time might be mind-dependent.

- Doubts about whether it is our consciousness that tells our body how to act have started to emerge in the last 30 years in neuroscience where the experimental evidence showed that *our consciousness only becomes aware of our actions, but does not initiate them* [73]. In 1983 Benjamin Libet and his colleagues performed experiments [74], which suggest that the consciousness does not initiate volitional acts since we become aware of them approximately two hundred milliseconds *after* they were initiated by unconscious processes in the brain. Recent experiments reported in *Nature Neuroscience* in 2008 appear to confirm Libet's results and demonstrate that our awareness of our own actions is even more delayed (up to *ten* seconds): "There has been a long controversy as to whether subjectively 'free' decisions are determined by brain activity ahead of time. We found that the outcome of a decision can be encoded in brain activity of prefrontal and parietal cortex up to 10 s before it enters awareness" [75].

It should be noted that one can still think of a materialistic interpretation of the consciousness which do not lead to a contradiction if the consciousness travels in spacetime. According to such an interpretation in the context of Weyl's idea, the consciousness might "operate" at a level whose dimensions could be greater than four, which would imply that spacetime is not applicable to all levels of reality. And indeed it does not appear realistic to expect that any *macroscopic* concept will be applicable to *all* scales of what exists. At some level lying 'beneath' the macroscopic level of our everyday experience, the properties of the world will inevitably be quite different from what we now know from our macroscopic experience; we have already started to observe such discrepancies at the quantum level. With this in mind, it is not completely unthinkable to expect that at some sub-microscopic level the whole histories of the elements at that level are not entirely given *at once*; in other words, what exists at that level is not a "frozen" world and therefore some kind of change could exist there. This makes it possible for the consciousness to "operate" at such a sub-microscopic scale.

The above difficulties with the nature of consciousness are not the only challenge to Weyl's explanation of the origin of our feeling that time flows. As we will see below it appears that Weyl's explanation faces more challenges and even appears to have some absurd consequences. At this point, I think we should have a clear view of the whole picture – as it does not appear realistic to hope that the space-time world view can be refuted, the only way to reconcile this view with the inter-subjective fact that we feel time flow seems to be Weyl's conjecture; we should either find another explanation or deal with its consequences (as long as they do not contradict experimentally-confirmed facts).

To demonstrate perhaps the most provoking consequences of Weyl's idea, imagine that you examine how you realize yourself. You are aware of your own body only at one constantly changing moment – the moment 'now'. According to Weyl, the reason for this is that our consciousness crawls upward along the worldtube of our body, realizing at each moment only a small region of the worldtube. The consciousness is always localized in an extremely small area of our worldtube which we perceive as our present body. This is not an additional assumption by Weyl – it merely reflects the inter-subjective fact that we realize ourselves only at the moment 'now' whose duration is unknown. A disturbing consequence from this fact is that the consciousness (the entity about which we know almost nothing) is to some extent *independent of our body*, since our past and future bodies exist as consciousnessless bodies – our consciousness is not there.

It cannot be assumed that our consciousness is spread along the whole worldtube of our body, because such an assumption leads to an obvious contradiction with the fact from our everyday experience – that we realize ourselves only at the moment 'now'. If our consciousness existed along the entire worldtube, i.e. if all our bodies – past, present, and future – possessed consciousness, then our life would be quite different: we would realize ourselves and the world at *all* moments of our life *at once* and there would be no flow of time; we would be in an eternal God-like state.

So, the feeling that time flows in the four-dimensional world is created by our consciousness which 'moves' toward the future part of the worldtube of our body, leaving our past bodies consciousnessless and 'giving life' to our also consciousnessless future bodies.

With this in mind we can now complete the explanation of the twin paradox. The twins exist at all events of their worldtubes, but each of them realizes himself only at a single event when his consciousness reaches and realizes that event. There should be no difference in the mind-dependent flow of time of the twins (i.e. in the advancement of the consciousness of each of them along his worldtube); at least there is no macroscopic reason that can cause any change in the time flow for one of the twins. Then when five years have passed for twin B and his consciousness reaches the event of the meeting M, he will be happy to meet his brother. However, this will be a very strange meeting – twin B will be meeting his brother from his brother's future. As twin A's consciousness moves in the same way as B's consciousness, five years have elapsed for A as well and his consciousness realizes event I; so his consciousness is five years behind B's consciousness. When twin A realizes event M, he will be meeting his brother from his brother's past; B's consciousness will be five years ahead at event N.

This provoking explanation of the twin paradox follows unavoidably from Weyl's idea. And that explanation has perhaps an even more provoking implication – can we be certain that some of the people we meet are not just consciousnessless bodies? For example, such a situation could occur when we meet people who travel a lot by plane, but their consciousness would be only a fraction of a second ahead of ours.

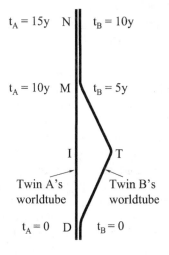

I have not explicitly stated it so far, but it is already evident that there is no free will in spacetime (i.e., in a four-dimensional world). As physical bodies we do not have any freedom to determine our actions since the worldtubes of our bodies are entirely given in spacetime (recall that the world-tube of a person's body contains the whole life of the person *given at once* like a film strip of an old movie contains the whole movie script at once). This worrying implication of the spacetime world view will

116

be discussed in the next chapter but here I would like to mention what appears to be another possible implication of Weyl's conjecture.

Weyl's explanation of the perception of time flow in spacetime contains very little information about the nature of consciousness – only (i) what is contained in the inter-subjective fact (that we are aware of a single moment of time which constantly changes toward the future), and (ii) what is necessary to explain the feeling of time flow correctly (that the localized consciousness "moves" toward the future part of a person's worldtube). As the consciousness is in some (still unknown) sense independent from our physical bodies (since our past and future bodies are consciousnessless), it is not unthinkable to assume that it might have some kind of freedom. For example, when we analyze the information from our senses stored in our brain and think of the different courses of action, our decision to act in a specific way might be different from that already realized in our worldtube. If such a situation occurs, we will be able to detect conflicts between our intentions and our actual actions, i.e. conflicts between our, to some extent, free consciousness and the predestined life of our body.

I think everyone can try to recall whether they had in their lives such cases. Friends, colleagues, and students have told me many strange stories which started to make some sense to them when they learned about Weyl's conjecture. I myself also had a number of very bizarre experiences which I can explain only as manifestations of conflicts between our to-some-extent free consciousness and our not free body. I think I can comfortably share two of those experiences with you.

When I was in my early teens my dream was to become a football player in the FC Botev Burgas (now PSFC Chernomorets Burgas) which competed in Bulgaria's top football league. But even then I knew very well that this was only a dream. However, one summer day when I was going to swim (Burgas is a large city on the Black Sea coast with nice beaches), I was stopped in the street by two men who told me they were coaches in FC Botev Burgas; they explained that the football team needed "several tall boys like you" and asked whether I would come to the stadium for a tryout the following day. I was overwhelmed with excitement and wanted to reply "Of course, I would be glad to come." I was completely shocked, therefore, when I heard myself say "I am sorry. I do not have time since my friends are waiting for me on the beach" and continued walking toward the sea. Immediately after leaving the stunned coaches, I said to myself "You are an idiot! Turn back right away, apologize, and say what you wanted to say." It was like in a nightmare – I simply could not do it since something was dragging me forward. Obviously, it had not been predestined that I become a football player.

Years later, while studying Physics at the University of Sofia, from time to time I helped support my studies by acting as a film extra. In one film about a political assassination, the extras had to play people walking in a small park in the downtown area of Sofia. On that day, I was dressed as a young officer and had to walk and talk with a young woman. In fact, she was doing the talking, passionately explaining her dream to become an actress. I was listening politely trying to understand her since the life of an actor had never attracted me – I was genuinely interested in my physics studies and in becoming a physicist because I believed that that was the only reliable way to understand the world. During a break, the film director approached us and I thought he wanted to speak to the young lady, but instead he told me "You have an interesting face. Would you like to try something more serious?" Instantly overcoming the surprise, I thought "This may not be a bad idea since I could earn more and buy the books I need rather than only those which I can afford." So I intended to reply "That is very tempting." To my absolute amazement and embarrassment, I told the film director "This is blackmail." I guess you can imagine how he looked at me and he quickly walked away. The young lady stared at me intensely and, visibly shaken, told me "You are totally crazy." The unanticipated prospect of my becoming an actor was brutally crushed through me, but not by me. But now I am truly happy that, at that time, I acted "totally crazy."

Some might say that Weyl's idea leads to absurd consequences. But it should not be brushed aside because of such a suspicion since those "absurd consequences" may turn out to be only apparent. His idea should be examined rigorously because, as we have seen above, it may very well turn out to be the only way to reconcile the spacetime world view and our undeniable feeling that time flows.

9 SPACETIME, FREE WILL AND ... THE MEANING OF LIFE

> Things are never quite the way they seem.
> S. Ridgway, Lyrics "Camouflage"

The debate over free will has been going on for centuries, but Minkowski's revolutionary view of reality as an absolute four-dimensional world (which *explained the profound meaning* of Einstein's discovery that observers in relative motion have *different* times) completely changed the essence and the rules of the free will debate. It turned out that the question on whether or not we possess free will *crucially* depends on the question of the nature of spacetime, i.e. on the dimensionality of the world. Free will may exist *only* in a three-dimensional world where the whole world and our three-dimensional bodies exist only at the present moment, which seems to imply that we might be free to determine our future because, on the presentist view, it does not exist in any sense and is therefore *undetermined* (but the very existence of the debate over free will before the advent of relativity and its spacetime formulation by Minkowski indicates that it is far from certain that free will can exist even in a three-dimensional world).

In the Minkowski four-dimensional world, however, the whole history in time of every macroscopic body is *existing at once (en bloc)* there as the body's worldtube, which means that there is no free will in a four-dimensional world. In Chapter 5 we saw that the relativistic effects length contraction, time dilation, and the twin paradox effect (and, most importantly, the *experiments* which confirmed them), would be impossible if the worldtubes of the (macroscopic) objects involved in them did not exist as four-dimensional objects. You recall that the impossibility of these relativistic experimental results in a

120

three-dimensional world was the obvious justification of the statement that the relativistic experimental evidence proved both the reality of the worldtubes of macroscopic bodies and the reality of spacetime (the Minkowski four-dimensional world).

Therefore, as the worldtube of our body contains our entire life (since our worldtube comprises our body which *exists equally at all moments of our life*), it does follow from the reality of the worldtube that we (as physical bodies) do not have any free will.[1] This can be best visualized by again using the analogy with a film strip of an old movie – we watch on the screen the actions of the main character but we know that her actions are completely predetermined since she is "doing" on the screen precisely what is already on the film strip (where her "worldtube" is entirely given).

It certainly may appear shocking that our life is entirely predetermined. Moreover, our daily experience captured in the so called "common sense" tells us the opposite – it is self-evident that such a ridiculous claim is wrong and everyone can refute it instantly and easily by acting in a way he or she *freely* chooses. The "common sense" reaction is indeed the most common reaction when people hear for the first time that the theory of relativity is a theory of a real four-dimensional world where there is no free will. Even physicists might react in this way. Years ago, I remember how a friend of mine (a physicist) reacted when we were discussing the implications of the Minkowski world during a long walk in a windy day along the Black Sea shore. When I told him that there is no free will in spacetime (i.e. in the four-dimensional world revealed by the theory of relativity), his reaction was almost furious in perfect harmony with the gusting wind – That is complete nonsense! My free actions are the strongest proof that I do have free will. I can do whatever I please. I can move my umbrella this way or that way or can choose not to move it all. When he saw that his "proof" was not impressing me, he started slowly to switch from emotion to reason. Then we discussed whether there existed, in principle, a way to prove through some actions that we have free will. The following scenario convinced my friend that that was impossible. Imagine watching an old movie in which the main character reacts upon reading that the theory of relativity rules out the

---

[1]Therefore any scientific study of free will should start with addressing the question "Is spacetime real?" or, equivalently, "What is the dimensionality of the world?" That is why, I suspect that a grant application for research on free will would have no chances if it did not indicate that the issue of the nature of spacetime would be an integral part of the research. For the same reason, I think it is clear that any (professional or general) discussion of free will, which does not even mention this question, risks to turn into insignificant chat.

existence of free will by saying "I do not care what that bizarre theory claims. I know I can do whatever I decide right now". We can only smile at the genuine confidence of the main character since we know that what is being projected on the screen is predetermined since it is entirely given on the film strip.

Throughout this book we have seen that we should be very alert whenever we encounter any apparent contradictions between "common sense" and new scientific results. The history of science is full of such apparent contradictions and in every single case it was "common sense" that was at fault. Recall only several examples. It had been against "common sense" that the Earth is a sphere (the view that the Earth was flat turned out to be an illusion). It had been against "common sense" that the Earth was orbiting the Sun (the belief that the Earth was motionless turned out to be an illusion). It had been against "common sense" that a free body (e.g. a projectile) can move without a mover (it turned out to be an illusion that an uniformly moving body needed a mover). It will be very stimulating and helpful (especially to those who tend to resist changes to their worldview implied by science) to try to continue this list by giving as many such examples as you can.

As seen from the topics discussed in the book, it is sometimes inexplicably difficult to overcome the "common sense" resistance to counter-intuitive scientific results. This is especially true in the case of the implications of the theory of relativity and particularly of its spacetime formulation given by Minkowski. In a popular book on relativity by the great Russian physicist Landau and Rumer – Что такое теория относительности (What is the theory of relativity) – the authors illustrated the clash of "common sense" with the real facts by a well-known anecdote about a farmer who exclaims in disbelief "This can't be!" (Этого не может быть!) when at the zoo he sees a giraffe for the first time [76].

This can't be! (Source: [76])

I know, it is easy to say "Do not trust "common sense"!" but how can one accept such a provoking statement that we do not have free will? I think the honest way of dealing with it is to follow these steps:

122

Step I The major step – rigorously examine the arguments that the relativistic experimental evidence proved the reality of spacetime (i.e. the four-dimensionality of the world). These arguments are summarized at the end of Chapter 1 and in Chapter 5. I would suggest to start with the thought experiment in Chapter 5 of two relatively moving meter sticks with cameras and changing lights, which visualizes real experiments of length contraction, since it can be certainly understood by everyone. As this experiment is also a visualization of Minkowski's explanation of length contraction, everyone can understand his arguments that the theory of relativity is a theory of an absolute four-dimensional world (spacetime).[2]

Step II Scrutinize our belief that we do act according to our intentions. Do we have any means to prove that it is we who are the masters of our actions? Then examine rigorously Libet's and more recent experiments in neuroscience, briefly discussed in the previous chapter, which imply that our consciousness does not appear to be calling the shots since we become aware of our actions only *after* they have been initiated by unconscious processes.

Step III Examine Weyl's explanation of our feeling that time flows is caused by our consciousness, which moves along our worldtube toward the future events and at every moment "reads" the information from our senses stored in our brain. As we saw in the previous chapter Libet's and the recent experiments in neuroscience are in full agreement with this mind-dependent explanation of the flow of time. Reflect on the simple question – How do we know that Weyl's conjecture is not the accurate explanation of our feeling of time flow? Do we have any means to rule out that explanation? Then you may realize that it is only the theory of relativity which tells us something relevant and meaningful about such a mind-dependent explanation of the flow of time (and you will see why Weyl went ahead with this explanation that was first contemplated by the Eleatics, Aristotle and Augustine as we saw in Chapter 2). This may convince you to consider seriously the possibility that our consciousness is not

---

[2]The implication of the spacetime world view that we do not have free will is almost certainly behind the fact that the arguments (based on the relativistic experimental evidence), which prove the reality of spacetime, are often merely ignored (they have never been refuted). I find such an attitude irrational since by refusing to face the fact that the world is four-dimensional will not make it disappear and leave behind the familiar evolving in time three-dimensional world.

telling our body how to act since it is only a passive observer of our material life.

Step IV If at the end it does follow that there is no free will, be honest and say – if the world is such, then what? What is the meaning of all this? At least out of curiosity try to examine such a four-dimensional world in which no free will exists.

As on the spacetime world view Weyl's conjecture is the only explanation of the inter-subjective phenomenon of time flow – that it is the consciousness which creates our feeling of a flowing time – then there is no contradiction between the static four-dimensional world of relativity and our *feeling* that it is the consciousness that decides what our future actions will be – the consciousness only "reads" the information about our actions stored in the brain at the consecutive moments of time but incorrectly interprets that fact in the sense that it is the consciousness which initiates our actions and because of that we have free will.

All this may make sense *logically*, but how can one accept such a drastic shift in our understanding of the world and our place in it? An option is to act as before hearing about the implications of the theory of relativity, i.e. by merely ignoring the whole issue of the nature of reality (and, of course, this book as well). In his book *The 7 Laws of Magical Thinking: How Irrational Beliefs Keep Us Happy, Healthy, and Sane*, Matthew Hutson calls such an approach to uncomfortable facts "magical thinking" – "magical thinking . . . offers a sense of control and a sense of meaning, making life richer, more comprehensible, and less scary" [17, p. 6].

I have written this book because I think we all as rational and intellectually curious beings should not be afraid to confront the truth about the nature of reality, no matter how counter-intuitive it may be. For this reason, let me outline how I try to understand the issue of free will in spacetime.

In innumerable discussions of the non-existence of free will in spacetime with both experts and non-experts, one of the most difficult questions is about how to deal with its apparently huge implications for virtually every aspect of our lives. As it would take a whole book to describe and analyze all those implications properly, I will briefly discuss only several no-free-will-related questions.

The most often worry is that when people accept that they have no free will, they will either lose motivation for working hard to achieve significant results in their lives, or will get depressed that nothing depends on them. I guess, you have already detected that there is something wrong with this worry. Indeed, it presupposes that we have free will to act differently when we learn that we do not have free will. No free will means precisely that – that nothing depends on us and therefore we are neither free to lose motivation nor free to get depressed.

If you reflect a bit on this situation, you may be pleasantly surprised to see it from a completely different angle. Accepting that we do not have free will may make life truly comprehensible and less scary than if we hide from the challenges of modern science by employing "magical thinking", for example. As we know that nothing depends on us we can adopt an attitude of "observers" of our own life. Then we can no less genuinely enjoy our achievements since we "as if participated" in all efforts leading toward them. We should not blame ourselves for our failures (since all is predetermined) and therefore we should not get depressed if, of course, our worldtube is depression-free after the event of the realization that free will does not exist in spacetime. Like anyone else, my wife, our son, and I also had difficult moments in our lives but this "spacetime philosophy" has kept "us happy, healthy, and sane" (undoubtedly, a manifestation of what is on our worldtubes after the event we realized the implications of the spacetime worldview for our free will).

Sometimes we hear wise people say "Don't judge!" Most probably, they have in mind the fact that in many cases we are inclined to judge others without knowing the whole story. But they may also mean that we, mortals, are not empowered to make true judgments and only a supreme being can do that. The "spacetime philosophy" offers another understanding of that advice – in this moral meaning of the term "judge" we should not indeed judge anyone since we all *only experience* our "life scripts" (our lives) which are realized *en bloc* in our worldtubes (meaning that we are not masters of our actions). How wise this "Don't judge!" in the spacetime context is, can be illustrated, for example, in the case of people who break some rules in society. If we judge them (in the moral sense), they would be, in a sense, doubly punished – first, by their bad luck to be "assigned" such a worldtube (i.e. a "life script") containing some rule-breaking (e.g. speeding on the highway) for which they are held responsible (which is part of their "life script"), and, second, by our moral judgment.

The most difficult no-free-will-related question appears to be about the meaning of life – what is the meaning of our existence if nothing depends on us? Never before the advent of the theory of relativity (and particularly its spacetime formulation by Minkowski) had the perennial question of the meaning of life been linked to science. This question has always been regarded as belonging to literature, philosophy, religion, but not to science.

Now a major consequence of the theory of relativity implies that we do not have free will, which in turns appears to suggest that our existence is meaningless. Here again we have a situation where all options (including purely logical) should be examined. It is quite evident that, in fact, it is our *physical bodies*, whose life is predetermined. As we saw in the previous chapter our consciousness is in some sense free from the body because it "moves" toward the future part of the worldtube of our body, leaving our past bodies consciousnessless and "giving life" to our also consciousnessless future bodies. An obvious logical option here is that the meaning of life may refer to the consciousness, not to the physical body. But such an interpretation crucially depends on an open question – what happens to the consciousness when it reaches the end of the worldtube of one's body. Unfortunately, this question cannot be answered on the basis of the two facts that led Weyl to the mind-dependent view of time flow – that (i) special relativity reveals that the world is four-dimensional and (ii) we realize the world and ourselves at the present moment. These facts do not provide even a hint of what happens to the consciousness at the end of the worldtube.

Now we could say either "We have analyzed all logical implications of the spacetime world view for our lives (particularly for free will, the nature of consciousness, and the meaning of life) and have nothing more to add" or "Let us see whether there exist some relevant and serious studies of consciousness which might shed some light on what happens to it when it reaches the end of a human body's worldtube." Excluding religious and philosophical writings about consciousness (or the soul), the only reliable work on whether or not consciousness is an independent entity appears to be a recent *scientific* study [77] of what is called near-death experience or out of body experience [78], [79].

Such experiences are reported by people who have been very close to death, usually after they have been pronounced clinically dead. Not all who return to life after clinical death report (or are willing to report) such experiences, but due to recent advancements in the techniques of cardiac resuscitation, more and more people come forward with strikingly similar stories. What is common in these phenomena is that people recall looking down from above at their bodies in the operating room and at the medical staff there, then seeing a tunnel and bright light, or a "being of light." Some people also report seeing objects on the shelves in the room that can be seen only from above.

There have been a number of studies claiming that "Taken together, the scientific experience suggests that all aspects of near-death experience have a neuro-physiological or psychological basis" [80] as the title of an article published in 2011 states "There is nothing paranormal about near-death experiences: how neuroscience can explain seeing bright lights, meeting the dead, or being convinced you are one of them" [80]. The most recent (2013) study with "rats undergoing experimental cardiac arrest" also tried to explain near-death experiences with a "Surge of neurophysiological coherence and connectivity in the dying brain" [81]:

> High-frequency neurophysiological activity in the near-death state exceeded levels found during the conscious waking state. These data demonstrate that the mammalian brain can, albeit paradoxically, generate neural correlates of heightened conscious processing at near-death.

However, not all neuroscientists think that the phenomenon of near-death experience (NDE) has been explained. An example is an article also published in 2013 [82]:

> The present study showed that NDE memories contained more characteristics than real event memories and coma memories. Thus, this suggests that they cannot be considered as imagined event memories. On the contrary, their physiological origins could lead them to be really perceived although not lived in the reality. Further work is needed to better understand this phenomenon.

The most serious problem with all claims that the near-death experiences are generated by neurological processes in the dying brain is the failure to explain how a patient can see his or her own body from above. Even more inexplicable are the reports of some near-death survivors that they saw objects on the shelves in the room, where their bodies were lying, which can be seen only from above.

On 18 September 2008 BBC News reported about a scientific study headed by Dr Sam Parnia (University of Southampton) to verify these difficult-to-explain reports [77]:

> A large study is to examine near-death experiences in cardiac arrest patients. Doctors at 25 UK and US hospitals will study 1,500 survivors to see if people with no heartbeat or brain activity can have "out of body" experiences. ... The study, due to take three years and coordinated by Southampton University, will include placing on shelves images that could only be seen from above.

Dr Parnia told BBC that all reports about the near-death phenomenon should be properly examined: "This is a mystery that we can now subject to scientific study" [77]. He explained that the most important part of the study would involve setting up special shelving in resuscitation areas. There will be pictures on the shelves which, however, can be seen only from the ceiling. Dr Parnia did not hide the importance of the study – "If you can demonstrate that consciousness continues after the brain switches off, it allows for the possibility that the consciousness is a separate entity" [77].

The study, now called AWARE (AWAreness during REsuscitation), took more than the initially planned three years and was turned into "a long-term project that aims to study the relationship between mind consciousness and brain in patients who undergo cardiac arrest and clinical death" [83]. The latest AWARE Study Update on their website was in January 2013 [83]:

> The AWARE investigators have explained that owing to the exploratory nature of this study they do not anticipate there to be an end in the near future. Instead the study is likely to evolve into further research projects downstream with time. They are pleased to report the study is progressing well but have indicated that the results so far suggest more data and larger scale studies may be required. At this time, they anticipate being able to release the preliminary results obtained during the first five years of the study in September or October 2013 to mark the fifth anniversary of the launch of the study. This will be done through the appropriate scientific channels such as publications in scientific journals and possibly by means of a lecture, symposium or conference at a suitable venue if there is sufficient public interest. This would allow the data and results to be discussed in further detail.

In the meantime, Dr Parnia co-authored a book with Dr Young *Erasing Death: The Science That Is Rewriting the Boundaries Between Life and Death* (2013) [84], following his earlier book *What Happens When We Die?: A Groundbreaking Study into the Nature of Life and Death* (2006) [85].

After the announcement of the AWARE preliminary results in the Fall of 2013 we will be in a better position to say whether consciousness may survive the end of the worldtube of a human body and whether the meaning of life may indeed refer not to our physical bodies, but to our consciousness.

0 INSTEAD OF CONCLUSION: A FINAL CHALLENGE OF MINKOWSKI STRANGE WORLD

By now you have certainly realized most of the implications of Minkowski strange world and I guess you might tend to regard the implication that we, as physical bodies, do not possess free will as the greatest challenge. Indeed, that implication itself has a lot of worrying consequences. But there are also other consequences of Minkowski world, especially one, that appears to be even inconceivable. Now, instead of conclusion, I will discuss briefly that final and seemingly inconceivable implication which raises a different kind of worrying issues.

The presentist view of the world pictures it as an evolving universe which, according to science, does not need a creator. On the spacetime view, however, the world is a block universe – the entire history of the world is given *at once (as a whole – en bloc)* since all moments of time exist as the "points" of the fourth dimension of the world (exactly as all points of any of the three spatial dimensions exist at once). It may not be seen immediately, but *a block universe implies that it was created.* Probably the fastest way to realize this disturbing implication is by using again the analogy with the film strip of an old movie. The 'movie script' is *entirely given (at once)* on the film strip. The history of the world is *entirely given* in the block universe.

To see the challenge more clearly, consider the life of a single person, e.g. a very creative and productive movie script writer. Her entire life, with the smallest and even insignificant details, is given at once in the block universe. I doubt that anyone would be able to argue seriously that the 'script' of her life – her four-dimensional worldtube containing all events in her life (coordinated with the events of other people's lives), including the elaborated ideas of her fantastic movie scripts – occurred spontaneously. First, nothing can occur or happen in a block universe since time is entirely given there. Second, it looks utterly inexplicable how the enormous complexity captured even in a single person's worldtube can come into existence spontaneously. I think just one element of the movie script writer's life is sufficient to rule out the idea that her entire life occurred spontaneously – the very existence of *intentions* in her actions (although not all intentions are realized) revealed at later events of her worldtube. That is, how can a worldtube occur spontaneously given the fact that each part of the worldtube (corresponding to a given moment of time) contains intentions about its parts corresponding to later moments? It appears logical to assume that the gigantic 'world script' (the entire history of the universe) must be created. Perhaps taking seriously this assumption fully reveals what a challenge the spacetime view of the world is. I think it is sufficient to mention only two questions – who created it and how can the block universe be created given the fact that all moments of time are given at once. It is true that, formally, a block universe can be created in a second time. As the existing experimental evidence does not provide even the slightest hint of another time, I prefer to stop here. Moreover, there exists an unwritten suspicion shared by scientists that we may be asking a lot of wrong questions about the most fundamental features of reality. I will only remark that creationists have no chance against evolutionary biology, but a block universe can give them some food for thought.

I think it is worth stressing again in these concluding remarks that any rejection of the spacetime view of the world by using its apparently unacceptable implications as a pretext is baseless. Instead, such implications should be seen as what they actually are – a challenge that should be confronted by thoroughly scrutinizing the arguments behind the spacetime view of the world (starting with Minkowski's own arguments) by scientists, philosophers, and anyone willing to participate in such an industrious endeavor.

I know that some readers without a strong background in science may tend to follow the easiest way of ignoring any scientific arguments about the nature of reality. In such cases I would like to ask you to try to follow the scientific method whenever you are about to say "Look at all these troubling implications! The world definitely cannot be such a 'frozen' four-dimensional Universe!" It has been strongly and repeatedly emphasized in the book that the spacetime view of the world has been based solely on facts rigorously and unambiguously deduced from experiment. So, if you dislike some implications of the modern world view, you will be more motivated to try to find problems with it with the ultimate hope that you may succeed in disproving it. As arguments in science are faced and addressed (not merely ignored) the spacetime view of the world can be disproved *only* by refuting the arguments that both the theory of relativity and, most importantly, the relativistic experimental evidence are impossible in a three-dimensional world. As indicated at the end of Chapter 1 those arguments can be understood even by non-experts and therefore non-experts can try to refute them too. I am reasonably sure that any attempt to refute the arguments proving the spacetime world view will fail, but nevertheless I strongly encourage you to try. The best way to understand a view or an argument is to try to disprove it.

It is my hope that the worrying implications of the spacetime view of the world will not tempt anyone to take refuge from the blinding light of truth (revealed by modern physics) back into the deceivingly safe and comfortable cave of ignorance. We all know that such an option is not a solution to anything – the challenge will not disappear if we hide from it. The adequate and dignified way of dealing with all counter-intuitive views deduced from scientific achievements, particularly those that concern reality, and therefore affect everyone, is to face them by rigorously examining the arguments which support them. The rule is simple – refute the arguments or deal with the implications. We have nothing to lose. On the contrary – such an open-minded attitude toward challenging scientific discoveries will widely open our intellectual eyes and will enable us to see beyond illusions.

APPENDIX

Science College students who attended the course SCOL 270 (September 2008 – April 2009). Picture taken by Lima Kayello (Science College) on Monday, March 23, 2009

Liberal Arts College students who attended the course LBLC 397 (January 2009 – April 2009). Picture taken by Collin Potter Bonar (Liberal Arts College) on Wednesday, April 22, 2009

For over ten years most of the topics in this book have been covered in two courses at two Colleges at Concordia University – the Science College and the Liberal Arts College. At an advanced level those topics have been discussed in the two-semester course SCOL 270 attended by science students, whereas a concise and adjusted to non-science students version of the topics have been included in the one-semester course LBLC 397 given at the Liberal Arts College. A number of these topics were also included as part of the course PHIL 328 (on space and time) given at Concordia University's Department of Philosophy. I am grateful to all students who attended these courses and who contributed significantly through their questions and participation in the discussions to the appearance of this book.

REFERENCES

[1] Saint Augustine, *The Confessions*. In: *Great Books of the Western World*, Vol. 16, ed. by M.J. Adler (Encyclopedia Britannica, Chicago 1993), Book XI

[2] Martin Gardner (ed.), *Great Essays in Science* (Prometheus Books, Amherst, 1994) p. 48

[3] Plato, *The Republic*, Vol. II Transl. by P. Shorey (Harvard University Press, Cambridge 1942) Book VII p. 123

[4] J. Trefil, *The Nature of Science: An A-Z Guide to the Laws and Principles Governing Our Universe* (Houghton Mifflin Harcourt, Boston 2003) p. 98

[5] M. Planck, *Eight Lectures On Theoretical Physics Delivered at Columbia University in 1909*, Translated by A. P. Wills (Columbia University Press, New York 1915) pp. 129–130

[6] L. Smolin, *Time Reborn: From the Crisis in Physics to the Future of the Universe* (Alfred A. Knopf, Toronto 2013)

[7] G.F.R. Ellis: Physics in the Real Universe: Time and Space-Time. In [10] pp. 49–79

[8] J. Christian: Absolute Being versus Relative Becoming. In [10] pp. 163–195, Arxiv:gr-qc/0610049

[9] R.D. Sorkin: Relativity theory does not imply that the future already exists. In [10] pp. 153–161, Arxiv:gr-qc/0703098

[10] V. Petkov (ed.): *Relativity and the Dimensionality of the World* (Springer, Berlin 2007)

[11] G.F.R. Ellis, Comments on the webpage of his essay participating in the 2008 FQXi Essay Contest *The Nature of Time* (http://fqxi.org/community/forum/topic/361)

[12] H. Minkowski, Space and Time, new translation in [13].

[13] H. Minkowski, *Space and Time: Minkowski's Papers on Relativity*, edited by V. Petkov (Minkowski Institute Press, Montreal 2012).

[14] F. Weinert, *The March of Time: Evolving Conceptions of Time in the Light of Scientific Discoveries* (Springer, Heidelberg 2013) p. 166

[15] Quoted from: Michele Besso, *Wikipedia* (http://en.wikipedia.org/wiki/Michele_Besso

[16] K. Gödel: A Remark about the Relationship Between Relativity and Idealistic Philosophy. In: *Albert Einstein: Philosopher–Scientist*, ed. by P. Schilpp, 3rd. ed. (Open Court, La Salle 1988) p. 558

[17] M. Hutson, *The 7 Laws of Magical Thinking: How Irrational Beliefs Keep Us Happy, Healthy, and Sane* (Penguin, New York 2012)

[18] G. Nerlich, *Einstein's Genie: Spacetime out of the Bottle* (Minkowski Institute Press, Montreal 2013)

[19] D.F. Wallace, *Everything and More: A Compact History of Infinity* (Norton, New York 2003) p. 259

[20] A. H. Coxon, *The Fragments of Parmenides* (Parmenides Publishing, Athens 2009)

[21] W.K.C. Guthrie, *A History of Greek Philosophy, Vol. II: The Presocratic Tradition from Parmenides to Democritus* (Cambridge University Press, Cambridge 1969) p. 28

[22] J. Barnes, *The Presocratic Philosophers* (Routledge, London 1982) p. 127

[23] Plato, *Parmenides*, in: J.M. Cooper, *Plato: Complete Works* (Hackett Publishing Company, Indianapolis 1997) p. 362

[24] J. Barnes (Ed.) *Complete Works of Aristotle*, Vol. 1 (Princeton University Press, Princeton 1984)

[25] R. Moore: *Niels Bohr: The Man, His Science, and the World They Changed* (MIT Press, Michigan 1985) p. 196

[26] Saint Augustine, *Confessions* (Oxford University Press, New York 1998)

[27] Aristotle: *Physics*. In: *Great Books of the Western World*, Vol. 7, ed. by M.J. Adler (Encyclopedia Britannica, Chicago 1993)

[28] C.M. Linton, *From Eudoxus to Einstein: A History of Mathematical Astronomy* (Cambridge University Press, Cambridge 2004) pp. 38–45

[29] C. Ptolemy, *The Almagest*. In: *Great Books of the Western World*, Vol. 15, ed. by M.J. Adler (Encyclopedia Britannica, Chicago 1993) p. 12

[30] G. Galileo, *Dialogue Concerning the Two Chief World Systems – Ptolemaic and Copernican*, 2nd edn. (University of California Press, Berkeley 1967)

[31] A. Chalmers, "Galliean Relativity and Galileo's Relativity," in S. French and H. Kamminga (eds.), Correspondence, Invariance and Heuristics: Essays in Hounour of Heinz Post, (Springer, Heidelberg 1993) pp. 189-205.

[32] Aleksandre Koyré, *Études galiléennes: La loi de la chute des corps. Descartes et Galilée.* (Hermann, Paris 1939)

[33] V. Petkov, *Relativity and the Nature of Spacetime*, 2nd ed. (Springer, Heidelberg 2009)

[34] A. Pais, *Subtle Is the Lord: The Science and the Life of Albert Einstein* (Oxford University Press, Oxford 2005)

[35] A. Einstein, Autobiographical Notes. In: In: *Albert Einstein: Philosopher-Scientist*. P. A. Schilpp, ed., 3rd ed. (Open Court, Illinois 1969), pp. 1-94, p. 53

[36] M. Kaku, *Einstein's Cosmos: How Albert Einstein's Vision Transformed Our Understanding of Space and Time* (Atlas Books and W. W. Norton & Company, New York 2004) pp. 60–61

[37] A. Einstein, *The Collected Papers of Albert Einstein*, Vol. 2, transl. by A. Beck (Princeton University Press, Princeton 1989) p. 140

[38] A. Einstein, M. Grossmann, Entwurf einer verallgemeinerten Relativitätstheorie und einer Theorie der Gravitation. In: *Zeitschrift*

für Mathematik und Physik **62** (1913) S. 225–261. English translation "Outline of a Generalized Theory of Relativity and of a Theory of Gravitation" in *The Collected Papers of Albert Einstein, Volume 4: Swiss Years: Writings, 1912-1914* (Princeton University Press, Princeton 1995). A year later the collaboration between Einstein and Grossmann resulted in another paper [39] further clearing the path toward general relativity.

[39] A. Einstein, M. Grossmann, Kovarianzeigenschaften der Feldgleichungen der auf die verallgemeinerte Relativitätstheorie gegründeten Gravitationstheorie. In: *Zeitschrift für Mathematik und Physik* **63** (1914) S. 215–225.

[40] H. Minkowski, *Geometrie der Zahlen* (Teubner, Leipzig 1896)

[41] S. Walter, Minkowski, Mathematicians, and the Mathematical Theory of Relativity, in H. Goenner, J. Renn, J. Ritter, T. Sauer (eds.), *The Expanding Worlds of General Relativity*, Einstein Studies, volume 7, (Birkhäuser, Basel 1999) pp. 45-86, p. 46

[42] T. Damour, "What is missing from Minkowski's "Raum und Zeit" lecture", *Annalen der Physik* **17** No. 9-10 (2008), pp. 619-630, p. 626

[43] M. Born, *My Life: Recollections of a Nobel Laureate* (Scribner, New York 1978) p. 131

[44] H. Poincaré, Sur la dynamique de l'électron, *Rendiconti del Circolo matematico Rendiconti del Circolo di Palermo* **21** (1906) pp. 129-176.

[45] T. Damour, *Once Upon Einstein*, Translated by E. Novak (A. K. Peters, Wellesley 2006)

[46] H. Poincaré, *Mathematics and Science: Last Essays (Dernières Pensées)*, Translated by J.W. Bolduc (Dover, New York 1963) pp. 23-24

[47] H. Poincaré, *Science and Method*, In: *The Value of Science: Essential Writings of Henri Poincaré* (Modern Library, New York 2001) p. 438

[48] D. C. Dennett, *Darwin's Dangerous Idea: Evolution and the Meanings of Life* (Simon and Schuster, New York 1996) p. 21

[49] R. Geroch, *General Relativity: 1972 Lecture Notes* (Minkowski Institute Press, Montreal 2013), p. 7

[50] A.S. Eddington, The Relativity of Time, *Nature* **106** (1921) pp 802–804

[51] V. Petkov, *Inertia and Gravitation: From Aristotle's Natural Motion to Geodesic Worldlines in Curved Spacetime* (Minkowski Institute Press, Montreal 2012) Chap. 6 and Appendix C

[52] A. Sommerfeld, To Albert Einstein's Seventieth Birthday. In: *Albert Einstein: Philosopher-Scientist.* P. A. Schilpp, ed., 3rd ed. (Open Court, Illinois 1969) pp. 99-105, p. 102

[53] R.A. Mould, *Basic Relativity* (Springer, Berlin, Heidelberg, New York 1994), p. 83

[54] C.W. Misner, K.S. Thorne, J.A. Wheeler: *Gravitation* (Freeman, San Francisco 1973)

[55] R. d'Inverno: *Introducing Einstein's Relativity* (Clarendon Press, Oxford 1992), p. 33

[56] G.L. Naber: *The Geometry of Minkowski Spacetime* (Springer, Berlin, Heidelberg, New York 1992), p. 55

[57] R.P. Feynman, *The Character of Physical Law* (MIT Press, Massachusetts 1967)

[58] P. Grangier, G. Roger and A. Aspect, Experimental evidence for a photon anticorrelation effect on a beam splitter: a new light on single-photon interferences, *Europhys. Lett.* **1** (1986) pp. 173–179

[59] W. Rueckner and P. Titcomb, A lecture demonstration of single photon interference, *Am. J. Phys.* **64** (1996) pp. 184–188

[60] A. Tonomura, J. Endo, T. Matsuda, T. Kawasaki, and H. Exawa, Demonstration of single-electron buildup of an interference pattern, *Am. J. Phys.* **57** (1989) pp. 117–120

[61] G. Greenstein and A.G. Zajonc, *The Quantum Challenge* (Jones and Bartlett, Sudbury, Massachusetts 1997)

[62] R.P. Feynman, *QED: The Strange Theory of Light and Matter* (Princeton University Press, Princeton 1985) p. 10

[63] P.A.M. Dirac, *Principles of quantum mechanics*, 4ed. (Oxford University Press, Oxford 1958) p. 9

[64] A. H. Anastassov, Self-Contained Phase-Space Formulation of Quantum Mechanics as Statistics of Virtual Particles, *Annuaire de l'Universite de Sofia "St. Kliment Ohridski", Faculte de Physique* **81** (1993) pp. 135-163

[65] J. L. Synge, *Relativity: the general theory.* (Nord-Holand, Amsterdam 1960) p. 110

[66] W. Rindler, *Relativity: Special, General, and Cosmological* (Oxford University Press, Oxford 2001) p. 178

[67] A. Folsing, *Albert Einstein: A Biography* (Penguin Books, New York 1997)

[68] D. Brian, *Einstein: A Life* (Wiley, New York 1996)

[69] G.F.R. Ellis and R.M. Williams, *Flat and Curved Space Times* (Oxford University Press, Oxford 1988) p. 104.

[70] B. Greene, *The Fabric of the Cosmos: Space, Time, and the Texture of Reality* (Knopf, New York 2004) p. 142

[71] H. Weyl, *Philosophy of Mathematics and Natural Science* (Princeton University Press, Princeton 1949) p. 116

[72] E. Irvine, *Consciousness as a Scientific Concept: A Philosophy of Science Perspective* (Springer, Dordrecht 2013) p. 151

[73] B. Libet, *Mind Time: The Temporal Factor in Consciousness* (Harvard University Press, Cambridge 2004)

[74] B. Libet, Unconscious cerebral initiative and the role of conscious will in voluntary action, *Behavioral and Brain Sciences* **8** (1985) pp. 529–566

[75] C. S. Soon, M. Brass, H.-J. Heinze and J.-D. Haynes, Unconscious determinants of free decisions in the human brain, *Nature Neuroscience* **11** (2008) pp. 543–545

[76] Л.Д. Ландау, Ю.Б. Румер, *Что такое теория относительности*, 3-е доп. изд. (Советская Россия, Москва 1975) с. 41 [English publication: L. D. Landau, G. B. Rumer, *What Is Relativity* (Dover, New York 2003)]

[77] BBC News, *Study into near-death experiences* (18 September 2008); http://news.bbc.co.uk/2/hi/health/7621608.stm

[78] *Near-death experience*, Wikipedia, http://en.wikipedia.org/wiki/Near-death_experience

[79] C. Carter, *Science and the Near-Death Experience: How Consciousness Survives Death* (Inner Traditions, Rochester 2010)

[80] D. Mobbs and C. Watt, "There is nothing paranormal about near-death experiences: how neuroscience can explain seeing bright lights, meeting the dead, or being convinced you are one of them" *Trends in Cognitive Sciences*, Volume 15, Issue 10, pp. 447-449 (18 August 2011); http://www.cell.com/trends/cognitive-sciences/abstract/S1364-6613(11)00155-0

[81] J. Borjigina et al., "Surge of neurophysiological coherence and connectivity in the dying brain" *Proceedings of the National Academy of Sciences of the United States of America*, vol. 110 no. 35, pp. 14432-14437 (27 August 2013); http://www.pnas.org/content/110/35/14432.abstract

[82] M. Thonnard et al., "Characteristics of Near-Death Experiences Memories as Compared to Real and Imagined Events Memories" *PLoS ONE* (2013) 8(3): e57620; http://www.plosone.org/article/info%3Adoi%2F10.1371%2Fjournal.pone.0057620

[83] *Horizon Research Foundation*, http://www.horizonresearch.org/main_page.php?cat_id=279

[84] S. Parnia and J. Young, *Erasing Death: The Science That Is Rewriting the Boundaries Between Life and Death* (Harper Collins, New York 2013)

[85] S. Parnia, *What Happens When We Die?: A Groundbreaking Study into the Nature of Life and Death* (Hay House, Carlsbad 2006)

INDEX

About the author

Vesselin Petkov received a graduate degree in physics from Sofia University, a doctorate in philosophy from the Institute for Philosophical Research of the Bulgarian Academy of Sciences, and a doctorate in physics from Concordia University in Montreal. He taught at Sofia University and Concordia University, and also had a stint at the Physics Department of the Johannes Kepler University of Linz, Austria, before coming to Montreal in 1990.

Vesselin with his wife Svetla in Santa Lucia, Cuba in May 2009

He is one of the founding members of the *Institute for Foundational Studies "Hermann Minkowski"* (minkowskiinstitute.org) whose most distinct feature is the employment of a research strategy based on the successful methods behind the greatest discoveries in physics. In this sense the *Minkowski Institute* is without a counterpart in the world.

Image Credits

p. 2: Plato's Allegory of the cave – http://en.wikipedia.org/wiki/
File:Platon_Cave_Sanraedam_1604.jpg

p. 10: Satellite images of Montreal – Google Maps

p. 29: Raphael School of Athens: http://commons.wikimedia.org/
wiki/File:Raphael_School_of_Athens.jpg

p. 30: Parmenides: http://en.wikipedia.org/wiki/File:Parmenides.
jpg

p. 33: Zeno: http://en.wikipedia.org/wiki/File:Zeno_of_Elea_
Tibaldi_or_Carducci_Escorial.jpg

p. 34 Aristotle: http://commons.wikimedia.org/wiki/File:Aristotle_
Altemps_Inv8575.jpg

p. 36: Saint Augustin: http://en.wikipedia.org/wiki/File:Saint_
Augustine_Portrait.jpg

p. 41 Galileo: http://commons.wikimedia.org/wiki/File:Galileo.
arp.300pix.jpg

p. 55: Einstein: http://en.wikipedia.org/wiki/File:Einstein_
1921_by_F_Schmutzer.jpg

p. 65: Minkowski: http://en.wikipedia.org/wiki/File:De_Raum_
zeit_Minkowski_Bild.jpg

CPSIA information can be obtained
at www.ICGtesting.com
Printed in the USA
LVHW081541121120
671530LV00030B/1351